P9-DHL-670
CONTENTS

SAFETY FIRST

Though all the designs and methods in this book have been tested for safety, it is not possible to overstate the importance of using the safest construction methods possible. What follows are reminders; some do's and don'ts of basic carpentry. They are not substitutes for your own common sense.

- *Always* use caution, care, and good judgment when following the procedures described in this book.
- *Always* be sure that the electrical setup is safe; be sure that no circuit is overloaded and that all power tools and electrical outlets are properly grounded. Do not use power tools in wet locations.
- *Always* read container labels on paints, solvents, and other products; provide ventilation, and observe all other warnings.
- *Always* read the manufacturer's instructions for using a tool, especially the warnings.
- *Always* use hold-downs and push sticks whenever possible when working on a table saw. Avoid working short pieces if you can.
- *Always* remove the key from any drill chuck (portable or press) before starting the drill.
- *Always* pay deliberate attention to how a tool works so that you can avoid being injured.
- *Always* know the limitations of your tools. Do not try to force them to do what they were not designed to do.
- *Always* make sure that any adjustment is locked before proceeding. For example, always check the rip fence on a table saw or the bevel adjustment on a portable saw before starting to work.
- *Always* clamp small pieces firmly to a bench or other work surface when using a power tool on them.
- *Always* wear the appropriate rubber or work gloves when handling chemicals, moving or stacking lumber, or doing heavy construction.
- *Always* wear a disposable face mask when you create dust by sawing or sanding. Use a special filtering respirator when working with toxic substances and solvents.
- *Always* wear eye protection, especially when using power tools or striking metal on metal or concrete; a chip can fly off, for example, when chiseling concrete.
- *Always* be aware that there is seldom enough time for your body's reflexes to save you from injury from a power tool in a dangerous situation; everything happens too fast. Be *alert!*
- *Always* keep your hands away from the business ends of blades, cutters, and bits.
- *Always* hold a circular saw firmly, usually with both hands so that you know where they are.
- *Always* use a drill with an auxiliary handle to control the torque when large-size bits are used.
- *Always* check your local building codes when planning new construction. The codes are intended to protect public safety and should be observed to the letter.
- *Never* work with power tools when you are tired or under the influence of alcohol or drugs.
- *Never* cut tiny pieces of wood or pipe using a power saw. Cut small pieces off larger pieces.
- *Never* change a saw blade or a drill or router bit unless the power cord is unplugged. Do not depend on the switch being off; you might accidentally hit it.
- *Never* work in insufficient lighting.
- *Never* work while wearing loose clothing, hanging hair, open cuffs, or jewelry.
- *Never* work with dull tools. Have them sharpened, or learn how to sharpen them yourself.
- *Never* use a power tool on a workpiece—large or small—that is not firmly supported.
- *Never* saw a workpiece that spans a large distance between horses without close support on each side of the cut; the piece can bend, closing on and jamming the blade, causing saw kickback.
- *Never* support a workpiece from underneath with your leg or other part of your body when sawing.
- *Never* carry sharp or pointed tools, such as utility knives, awls, or chisels, in your pocket. If you want to carry such tools, use a special-purpose tool belt with leather pockets and holders.

Planning an Energy-Saving Strategy

Making your home more energy efficient will lower your fuel bills, and make you more comfortable indoors during the winter and summer. You may have to take action on several fronts: adding insulation, weather-stripping windows and doors, and filling up gaps in the walls and ceilings—just to mention a few. After the house has been tightened up, you will need to provide a controlled source of fresh air. This chapter helps you set priorities as you decide how to insulate, seal and ventilate your home.

Creating an Effective Weather Barrier

Few areas in the world are blessed with climates so mild that there is no need for protection from the heat and cold. Keeping a home comfortable throughout the year means preparing it to suit the climate. Those living in Minnesota may endure a few weeks of heat and humidity in July, but their main concern is keeping warm throughout the long, cold winters. For those living in Southern Florida, the situation is reversed; they have to deal with heat and humidity all year long.

Insulating a house cuts down on the flow of heat, keeping it outside in summer and inside in winter. Proper insulation results in lower heating and cooling bills.

The amount of time, money and effort required to make a home weathertight depends on the condition of the house, the degree to which it will be improved, and the amount of work to be done by the homeowner. A do-it-yourselfer can achieve wonders over the course of a couple of Saturdays by accomplishing simple, inexpensive projects, such as weatherstripping doors and windows and sealing up cracks. However, the job may call for more extensive measures, such as installing insulation.

In most cases, the cost of major remodeling eventually returns to you as savings in heating fuel accumulates. In general, the cost of insulation and weatherproofing is paid back in about seven years. During this time, however, those living in the house reap immediate benefits in the form of increased comfort.

Tracking the Travels of Heat

Heat always moves from a hotter surface to a colder one. In order to properly cut the flow of heat to or from a house, first select a strategy best suited to the path heat travels.

Conduction

An object feels hot or cold because heat is either passing from the object to your body or vice versa. This is called conduction. The same theory applies to a house on a winter night. The siding or brick on the outside can be very cold, while the wall surface inside is warm. The heat is passed, or conducted, through each material within the wall and eventually to the cold outside air. Conduction works the opposite way during the summer.

Controlling Conduction. To control the heat that is conducted into or out of a house, poor conductors must be placed between the interior and exterior of the house. Still air is a poor conductor, and therefore it makes a good insulator. That is why materials such as fiberglass and foam plastic are effective insulators. Fiberglass insulation contains glass fibers surrounded by dead airspaces, while foam plastic is made up of tiny isolated air bubbles.

Convection

The heat you feel when you put your hand above a candle comes from warm air rising. This is called convection. There are two ways in which a house loses heat to convection. First, when warm, inside air contacts a colder surface, such as a window

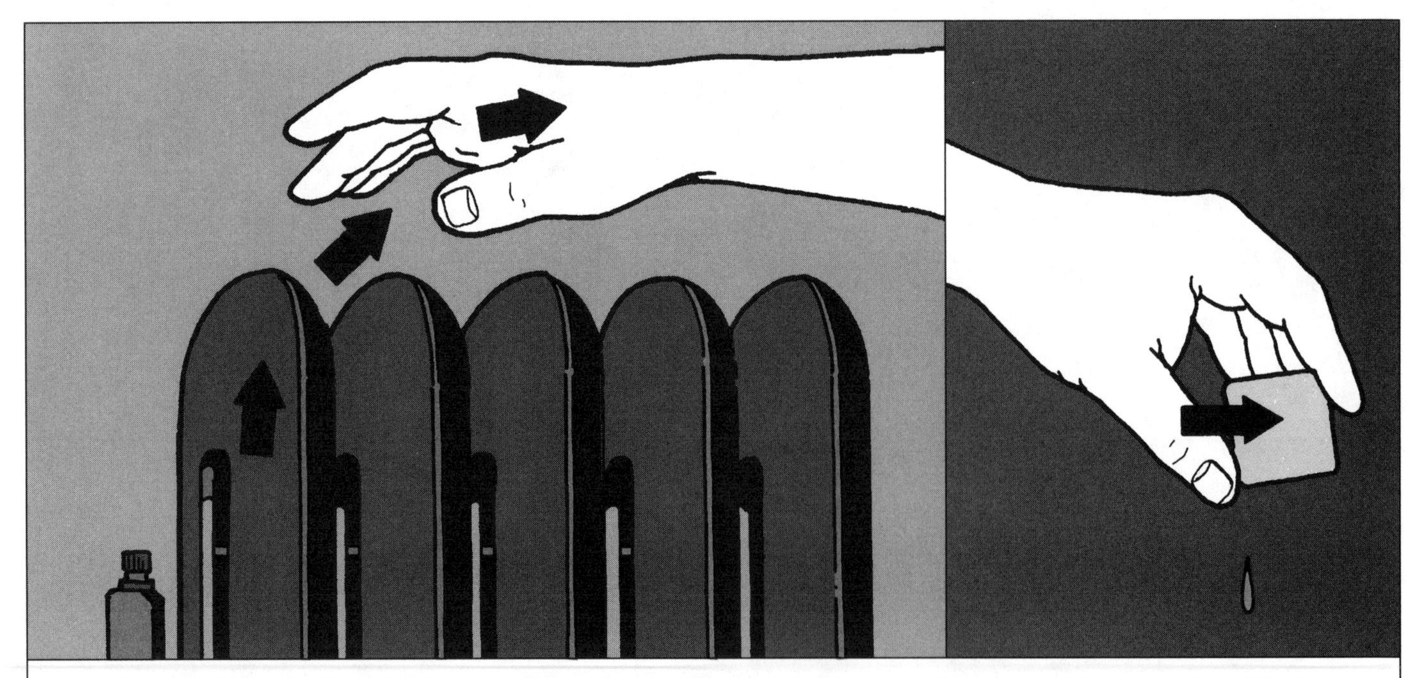

Conduction. Heat conducts from a warmer surface, such as a radiator, to a colder one, such as your hand (left). The direction of heat flow reverses when you pick up a colder object such as an ice cube (right).

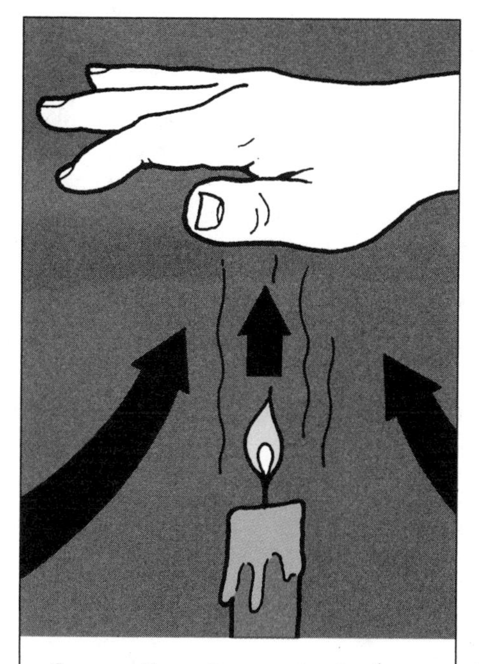

Convection. Convective heat is felt as warm moving air wafting above a hot source, such as a candle flame.

Controlling Convection. A house loses heat to convection in two ways: First, cold windows cool the inside air making it heavier and causing it to drop downward as a draft (left). Second, air finds a direct path to the outside through cracks and gaps in the structure (right). Replacing old windows with more efficient ones helps block both paths of convection. weatherstripping stops leaks that are caused by cracks in the existing windows.

or wall, the air cools and drops to the floor. It is then replaced with additional warm air. The second way is when cold air sneaks in through cracks. These problems result in uncomfortable drafts.

Controlling Convection. The convection that occurs around windows can be reduced by cutting down the heat that passes directly through the glass (via conduction), and plugging the cracks around the window sash. This is done by adding storm windows to the outside and by weatherstripping the windows. If the windows are in really bad shape replace them with more efficient units.

Radiation

A radiator or hot pan feels warm next to your hand. Heat waves from the hot radiator radiate to cooler objects. Like radio waves, this form of energy passes through the air without heating it. The energy becomes heat only after it strikes and is absorbed by a dense material. This form of heat transfer can be put to work to control heat gain or loss in a home. For example, the metal foil on some types of insulation reflects radiant heat back inside the house. However, this works only if metal foil is installed with an airspace separating it from the adjacent drywall or wall finish. The airspace stops the foil from simply conducting the heat energy through the wall materials. In hot climates foil is used in the attic to help cut the radiant heat that emanates from the roof.

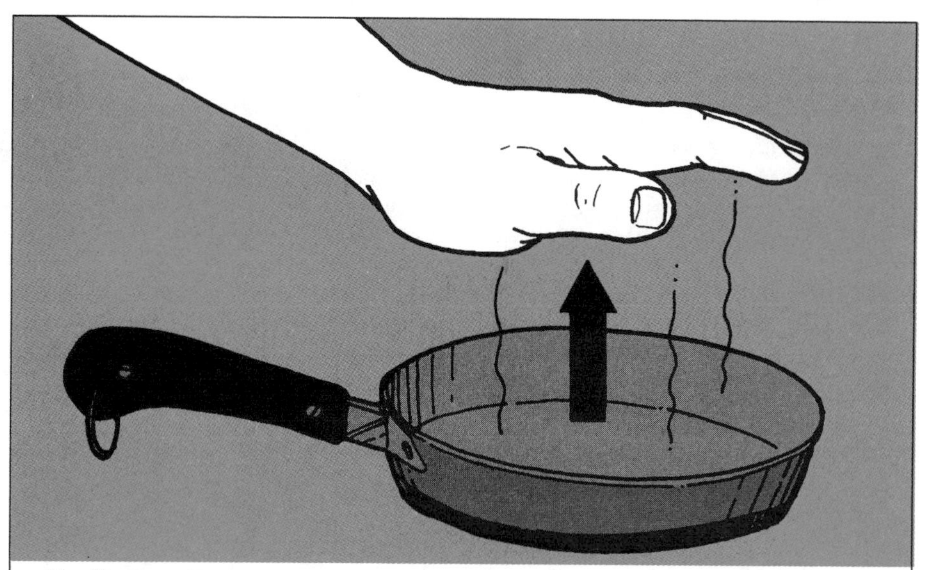

Radiation. Radiant energy does not heat the air. It only becomes heat when it reaches and is absorbed by a dense object. For example, heat from a hot pan reaches your hand held above it.

Getting the Most for Your Money

The key to getting the most from your insulating dollars is to prioritize. Find the weakest points in the home and fix them first. Then attack the next weakest points and so on. The trick is to figure out which areas are worse off than others. Uninsulated floors, walls and roof ceilings are best insulated before addressing other problems. After these problems are resolved, prioritizing becomes difficult. Questions arise: Should I add to the wall insulation or put more in the attic? Should I replace my old windows with more efficient ones?

Use the "Insulating Checklist" on page 9 to help make a decision. If you still do not know what to do, consider having an energy audit done on your home.

Getting Professional Help

Consider hiring a professional energy auditor to conduct a thorough examination of the heat loss that occurs in your home. An auditor's analysis can help determine the best places to spend your improvement dollars. To find an auditor, check in the yellow pages under headings such as "Energy Management" and "Energy Conservation Consultants." If the Yellow Pages do not help, contact the public utilities agency in your state to ask for references.

An energy auditor may use a variety of methods to assess the heat loss pattern in a home. Thermographic photography, for example, produces a picture that shows exactly where the heat is escaping. Another method uses a "blower door" to determine how much air passes freely from the outside to the inside. This is done by means of a powerful fan attached to a door that is temporarily installed in an outside doorway.

If the cost of a professional energy audit is more than your budget can spare, check to see if the electrical or gas utility company offers an energy evaluation. Many do, and while the service is not as extensive as a professional audit, usually it is free.

Targeting Insulation Levels

The effectiveness of thermal insulation is rated by measuring its resistance to heat flow and assigning it a number, called an R-value. The higher the number, the better the insulation.

The U.S. Department of Energy has established minimum insulation levels, based on climate and type of heating, for various areas of the country. The chart recommends higher R-values in attics and roofs in houses that are heated with electric resistance heat rather than other fuel types. This is because heat loss usually is greatest through the ceiling and electric resistance heat usually is the most expensive source. Use the map and chart as a guide to selecting target levels for insulating various parts of the home. The final decisions depend on whether or not you can actually remodel parts of your home to achieve these target levels.

Law of Diminishing Returns

Adding a little insulation to an uninsulated house saves a lot of heat and money. However, each doubling of that amount of insulation costs the same, and saves only half as much heat loss as the last doubling. This is because there is less and less heat to save.

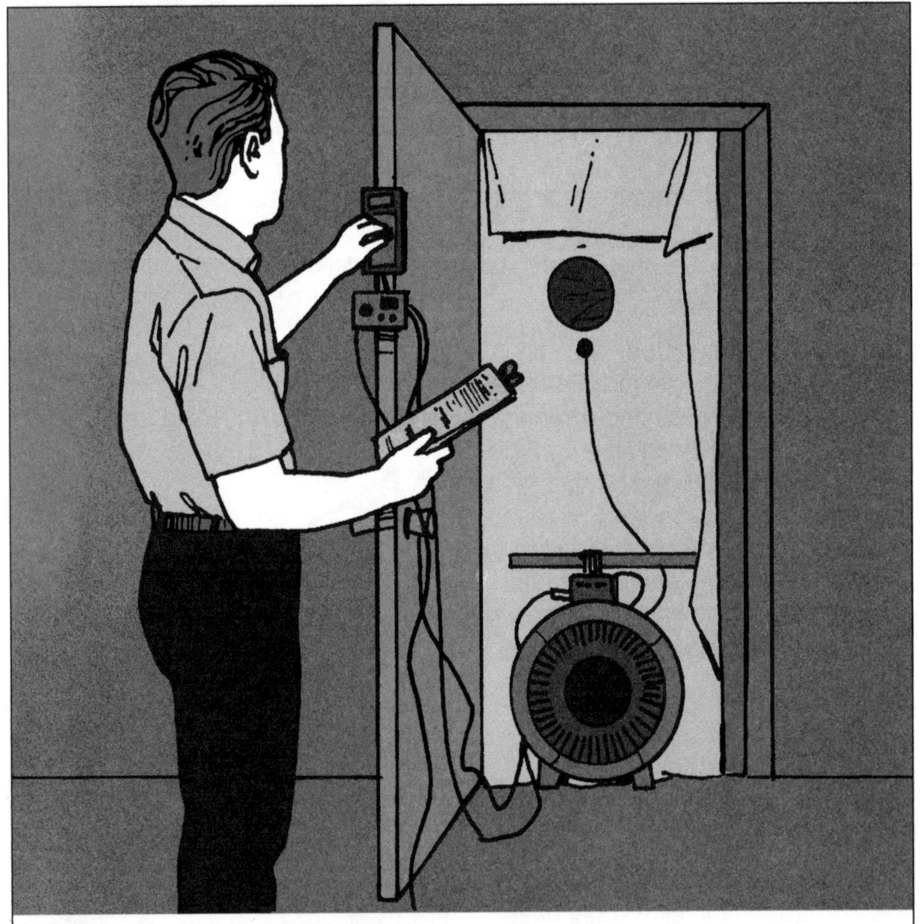

Getting Professional Help. One way to determine how badly your home suffers from leaks is to hire an energy consultant to run a test using a blower door. After temporarily mounting a powerful fan in an outside doorway, the technician de-pressurizes the house. The difference between inside and outside pressure determines the amount of air passing freely from the outside to the inside of the home.

Insulating Checklist

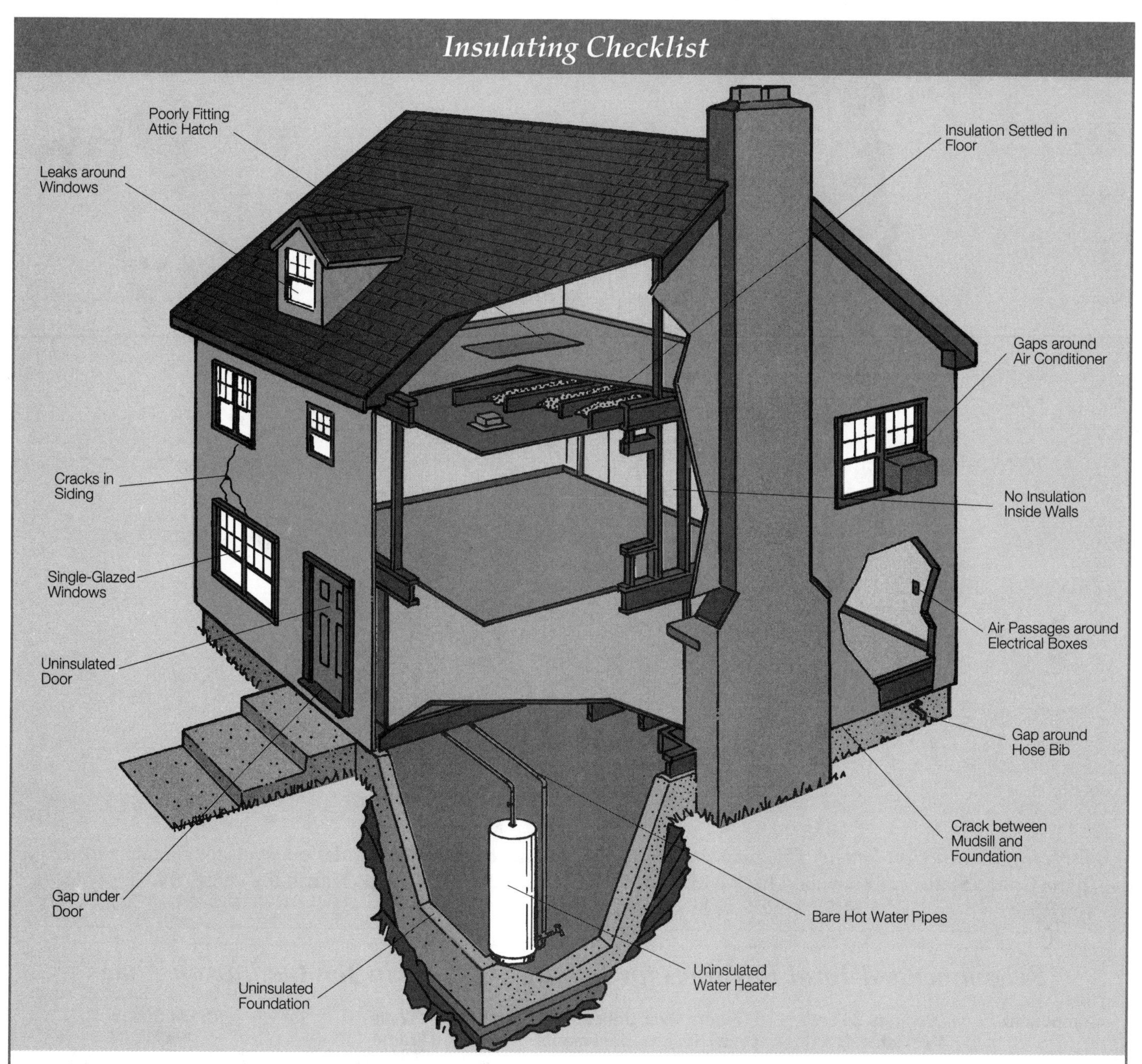

Ranked by likely cost, the following is a list of improvements from the lowest cost to the most expensive:

Wrapping Heater and Water Pipes. The single, most cost-effective thing to do is to wrap the water heater with an insulated blanket (sold at hardware stores). Then wrap all accessible hot water piping that runs from the water heater. Foam plastic sleeves, made for 1/2-inch diameter pipe, are available in 48-inch-long sections.

Sealing Major Air Leaks. Sealing leaky cracks around windows and doors is both economical and effective. It also results in a noticeable difference on cold winter nights.

Adding Insulation to the Roof or Attic. If the roof or attic presently has no insulation, make it a top priority. Topping off an underinsulated attic floor or roof also makes sense for those living in cold climates.

Adding Insulation to the Walls. Insulating uninsulated walls is next on the list of priorities. In the coldest climates, adding insulation to an underinsulated wall often makes economic sense, as does insulating concrete or masonry foundations.

Improving Inefficient Windows. Windows usually are the weakest weather barrier between the inside and the outside. If you live in a cold climate and do not have double-glazed windows (windows with two panes separated by an airspace) consider installing storm windows, or replace the old windows entirely with new, up-to-date units.

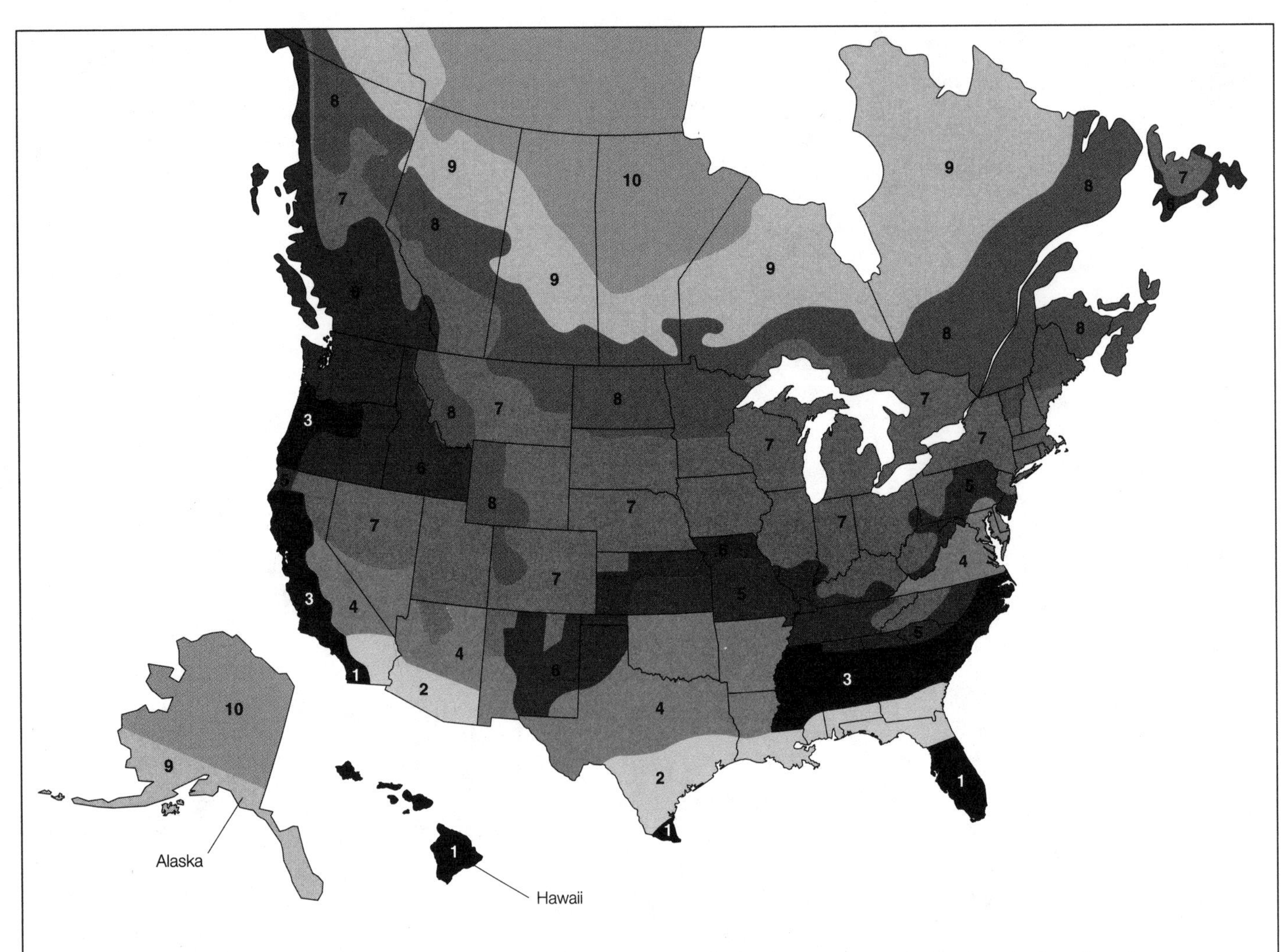

Targeting Insulation Levels. This map shows the recommended minimum insulation levels for climate zones in the United States and Canada. Check with your local building department to confirm the values that apply to your area.

Recommended Total R-Values for Existing Houses in Ten Insulation Zones[a]

Component	Ceilings Below Ventilated Attics		Floors Over Unheated Crawl Spaces, Basements		Exterior Walls[b] (Wood Frame)		Crawl Space Walls[c]	
Insulation	***Oil, Gas Heat Pump***	***Electric Resistance***	***Oil, Gas Heat Pump***	***Electric Resistance***	***Oil, Gas Heat Pump***	***Electric Resistance***	***Oil, Gas Heat Pump***	***Electric Resistance***
1	19	30	0	0	0	11	11	11
2	30	30	0	0	11	11	19	19
3	30	38	0	19	11	11	19	19
4	30	38	19	19	11	11	19	19
5	38	38	19	19	11	11	19	19
6	38	38	19	19	11	11	19	19
7	38	49	19	19	11	11	19	19
8	49	49	19	19	11	11	19	19
9	49	49	19	19	11	11	19	19
10	55	55	19	19	11	11	19	19

a These recommendations are based on the assumption that no structural modifications are needed to accommodate the added insulation.

b R-value of wall insulation, which is 3½ inches thick, will depend on material used. Range is R-11 to R-13. For new construction R-19 is recommended for exterior walls. Jamming an R-19 batt in a 3½-inch cavity will not yield R-19.

c Insulate crawl space walls only if the crawl space is dry all year, the floor above is not insulated, and all ventilation to the crawl space is blocked. A vapor barrier (e.g., 4- or 6-mil polyethylene film) should be installed on the ground to reduce moisture migration into the crawl space.

TOOLS & MATERIALS

Most insulation and ventilation jobs are easy to do and require simple tools. They are great do-it-yourself projects that bring immediate money savings. In fact, the hardest part of the job often is selecting the right materials. This chapter describes the tools you will need and helps determine the insulation product best for the job at hand.

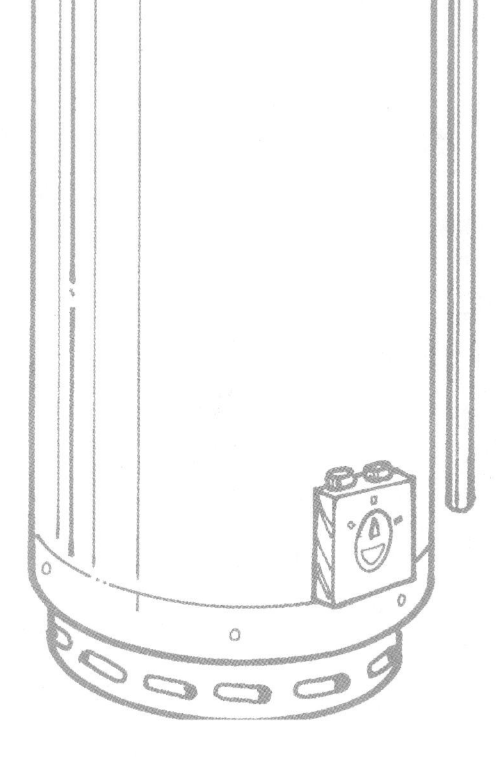

Tools

Active do-it-yourselfers will find that most of the tools needed to install insulation and weatherstripping are hanging on the wall of their garage or workshop. Special tools usually can be obtained from a hardware or home center store. In any case, the project and its necessary materials determine the tools needed, so do not buy them until these decisions have been made.

Bread Knife. A 10- to 12-inch-long serrated kitchen knife is great for cutting through blanket and batt insulation. Buy a cheap one or use an old one that will not be returned for kitchen use (after cutting insulation it

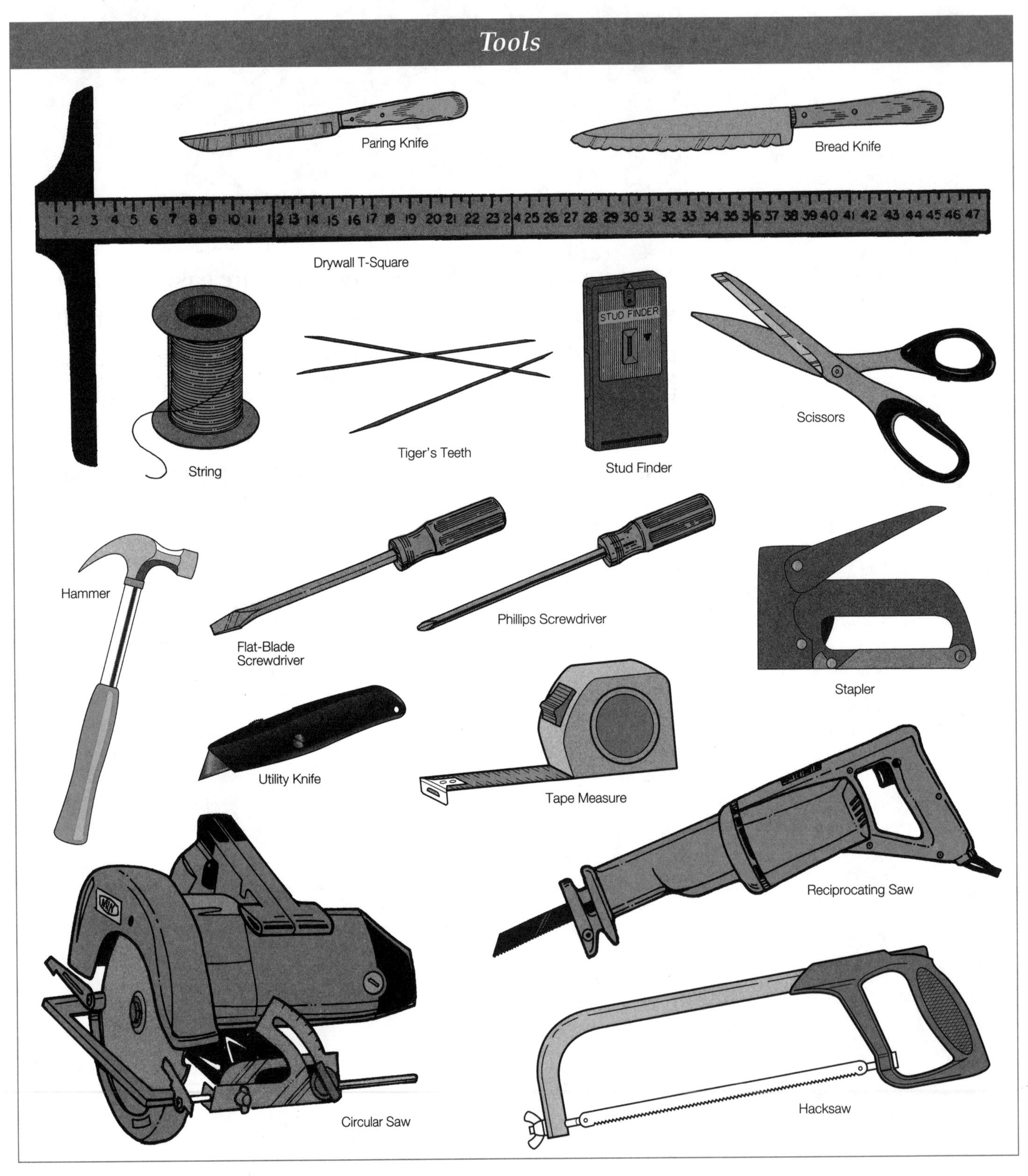

will be ruined). The knife becomes dull from cutting against a board and will require repeated sharpening (do not worry about sharpening each serration, simply run a file over both sides of the blade).

Circular Saw. A power circular saw is necessary for projects that involve frame carpentry.

Drywall T-Square. A 4-foot-long drywall T square is useful for measuring and marking rigid board. The aluminum edge doubles as a cutting guide. Consider purchasing this tool if you intend to work with rigid insulation. It can be used later for working with materials that come in 4x8-foot panels, such as drywall and plywood.

Hacksaw. This tool is needed to cut rigid weatherstripping and thresholds.

Hammer. A small tack hammer is perfect for hammering the small brads that come with some types of weatherstripping. A 16-ounce hammer is needed for bigger jobs, such as replacing windows.

Paring Knife. A utility knife blade is not long enough to cut through rigid foam. Something a little longer, such as a paring knife, does the trick.

Reciprocating Saw. This power saw comes in handy for major remodeling projects that require the removal of sections of wall (such as replacing windows and doors).

Screwdrivers. A basic tool kit is not complete without a flat-blade screwdriver and a Phillips screwdriver. They are used for many tasks including mounting storm panels and installing replacement windows.

Scissors. Scissors are useful for cutting housewrap and plastic sheeting.

Stapler. A stapler capable of driving staples up to 1/2-inch-long is necessary for installing blanket insulation.

Tape Measure. A 20- or 25-foot tape measure is adequate for general carpentry work including insulation and ventilation. Whether bought in the form of rolls or rigid lengths, weatherstripping materials have to be measured. Lengths of blankets and batts also must be measured.

Utility Knife. This inexpensive tool is used for numerous tasks, such as cutting insulating foams, tapes and air- and vapor-barrier sheet materials.

Stud Locator. A magnetic stud locator pins down the location of a stud by sensing the nails in the stud. A moving needle inside the stud locator indicates the stud's location. Unfortunately, nails do not always occur in the center of studs, so this type of device offers limited accuracy. A more accurate (and more expensive) device, called an electronic density detector, has a meter that indicates places of higher density which usually occur at the studs.

Protective Wear

Working with insulation materials may appear to be a harmless endeavor, since most likely, you will not have to work with power equipment. But cutting and installing fiber insulation releases fibers into the air, and these fibers can irritate your skin, eyes, nose and lungs. In addition, any time a knife is used to cut material, there is a risk of accidentally cutting yourself. It is important, therefore, to protect yourself by taking certain precautions and wearing protective gear.

Shielding your eyes, nose, hair, and other exposed areas of the body, ranks foremost when working with, or around, insulation. A pair of clear plastic goggles is a necessity. The most effective type come with wraparound pieces that fit snugly at the brow and temples. If you wear glasses, look for goggles that fit over the frames.

Next, it is important to protect your lungs. Disposable masks are appropriate for minor work such as stuffing fiberglass insulation into a ceiling from below. However, dustier projects, such as working with loose-fill insulation, and work done inside a confined space, such as an attic or crawl space, warrant investing in a permanent respirator fitted with replacement filter cartridges. Although wearing a respirator has some disadvantages (breathing is more labored and it gets warm inside the mask) the ability of the respirator to keep fibers and dust out of your lungs compensates for the temporary discomfort.

For additional protection wear a hat, such as a painter's hat, full-length trousers and a long-sleeved shirt. For extra insurance when working around fibrous materials, put an elastic band around your pants at each ankle,

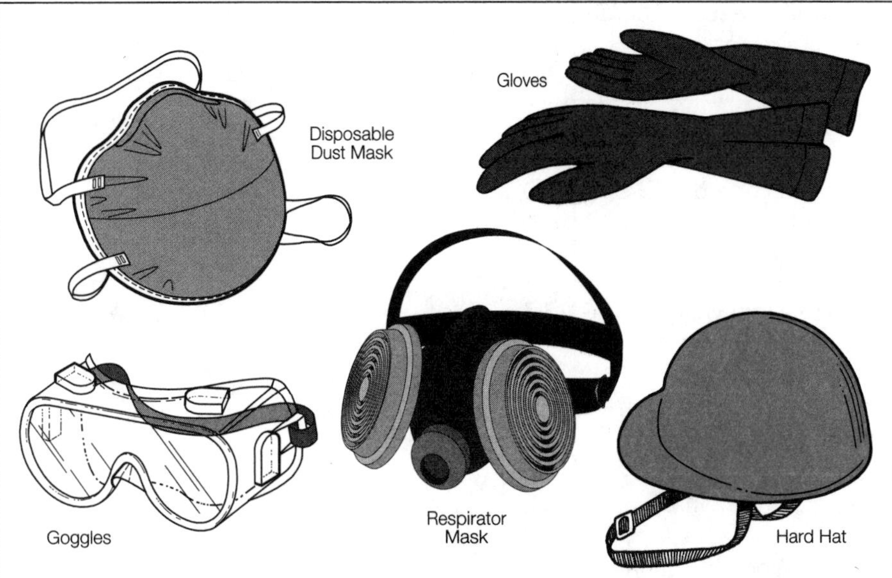

Protective Wear. It is important to protect your lungs, skin, and head when working with insulation.

and also around each sleeve at your wrists. If the work takes you into a space that has limited headroom, such as an attic or crawl space, wear a plastic hard hat to protect your head from the inevitable bumps against rafters or joists. Wear rubber gloves when working with pressurized foam sealants.

Choosing the Right Insulation

Begin by making a preliminary selection based on where the insulation is

Choosing the Right Insulation

INSULATION TYPE	R-Value Per Inch	Relative Cost Per R-value	Contains Vapor Barrier	Install It Yourself (DIY) or by Contractor (C)	APPLICATION: Cavities between Joists and Rafters	APPLICATION: Face of Framing or Over Existing Walls and Ceilings	APPLICATION: Foundations, Exterior	APPLICATION: Foundations, Interior	APPLICATION: Floor Slabs	Widths and Lengths
Blankets and Batts										
Fiberglass	3.3	Lowest	○[1]	DIY	●		●			15"xVar.[2] 23"xVar.[2]
Mineral Wool Batts	3.6	Lowest	○[1]	DIY	●		●			15"xVar.[2] 23"xVar.[2]
Rigid Insulation										
Phenolic Foam	8.5	Highest	●[1]	DIY		●	●	●	●	4'x8'
Polyurethane/Isocyanurate	7.2	Medium	●[1]	DIY		●	●	●	●	4'x8'
Polystyrene (extruded)	5.0	Medium		DIY		●	●	●	●	2'x8' 4'x8'
Polystyrene (beadboard)	4.0	Medium		DIY		●	●	●	●	2'x8' 4'x8'
Fiberglass board	4.0	Highest		DIY		●	●	●	●	4'x8'
Loose-Fill Insulation										
Cellulose (blown in)	3.7	Medium		C	●					
Cellulose (bagged)	3.7	Lowest		DIY	●					
Perlite (pellets)	2.7	Lowest		DIY	●					
Fiberglass (blown in)	2.2	Medium		C	●					
Mineral Wool (blown in)	2.9	Medium		C	●					

1. Foil facings can serve as a vapor barrier if all seams are taped—more practical with rigid, than blanket insulation.
2. Lengths of blankets and batts in a package vary according to the thickness of the insulation.

Insulation Facings

Blanket, batt and rigid insulation are available with or without a facing material. The tabs on kraft- or foil-faced blanket and batt insulation provide a surface for attaching the insulation to the framing, while the facing itself offers some protection against vapor passage. Kraft paper is only a marginal vapor barrier; foil facing is better. The weakest part of both are the sealed edges. The foil facing on rigid sheets can be taped easily, but adequately taping the many tabs and joints in blanket and batt insulation is a difficult job.

to be installed. Then narrow your choices by using other criteria such as cost per R-value, whether or not the insulation has an integral vapor barrier, and whether or not it can be installed by a do-it-yourselfer. The chart on page 14 compares the features of several of the most common types of insulation.These days, there is a perfect product for just about every insulating job. However, the availability of new polymeric foams, caulks and weatherstripping materials makes choosing one an overwhelming task in itself.

When choosing thermal insulation for a particular application, match its R-value to its use. Then consider the cost of the material and the demands of the job. If the form that the insulation comes in does not matter and there is enough space for any type of insulation, choose the product that most economically delivers the desired R-value. Unfortunately, choosing the right insulation usually is more complicated than this. Depending on the job, some types are better than others. It makes sense to understand the properties of each.

Blankets & Batts

Blanket insulation and batt insulation make up the lion's share of insulation installed by do-it-yourselfers. The only difference is that blankets come in long, single rolls, while batts come in shorter lengths packed several to a package. The choice is just a matter of what is most convenient for the job at hand. Both are available in thicknesses of 3 ½ to 12 inches and in widths of 15 and 23 inches, designed to fit snugly between wood framing spaced 16 or 24 inches on center. The materials used to make blankets and batts varies.

Fiberglass. Most homeowners think of fiberglass blankets and batts when they think of insulation. Also called glass wool or spun glass, this versatile insulating material consists of glass fibers held together with a binding substance. Fiberglass blanket and batt insulation is sold unfaced, or faced with kraft paper or foil on one side. The facing makes the insulation less permeable to air and water vapor and eases installation in some cases. The tabs along the edges provide an easy way to staple the insulation to the structural framing—a definite bonus for installing insulation overhead. However, for other applications, unfaced fiberglass is both cheaper and easier to install. Simply cut the pieces to the required length and place them between the studs or joists.

While fiberglass is not flammable, it melts when subjected to flame. The attached facing material, if any, is flammable. For this reason, fiberglass insulation that is used in walls and ceilings of occupied spaces must be protected by a noncombustible material such as drywall.

Mineral Wool. Sometimes called rock wool, this material is made of

Blankets & Batts. Flexible insulation comes in rolls (blankets) or pre-cut lengths (batts).

fibers spun from slag. It cakes when wet, losing most of its insulating value. The R-value of blanket and batt mineral wool insulation is about the same as fiberglass, but mineral wool blankets and batts are only available with a kraft paper facing.

Rigid Insulation

Made with various types of plastics, this insulation offers the highest R-value per inch of thickness. It also costs more than other forms of insulation. Use the requirements dictated by the task at hand to determine whether the higher cost is worth the advantages. Rigid insulation makes sense for applications in which keeping thickness to a minimum is desired (for example, when adding insulation to the room side of an underinsulated wall or ceiling). Unlike blanket insulation, rigid insulation can be applied to the outside of a foundation. It also can be nailed or glued to bare framing or to an enclosed wall. When faced with aluminum foil and with joints taped, rigid insulation also serves as a vapor barrier, eliminating the need for installing a separate polyethylene sheet.

Rigid boards are available in sheets that are 2 or 4 feet wide and 8 feet long. Thicknesses vary from 1/2 inch to 4 inches, though all types are not available in all thicknesses (3/4-, 1-,

1½-, and 2-inch thicknesses are the most common). Attain the desired R-value by adding more than one layer of whatever thickness is available. Building codes typically require foam plastic rigid insulation to be enclosed behind 1/2-inch drywall for fire protection.

Phenolic Foam. Phenolic foam, a pink extruded plastic, ranks as the most efficient, as well as the most expensive, type of rigid insulation. It also is highly resistant to fire, which makes it a good choice for lining the inside of walls and ceilings. Like most rigid insulation, phenolic foam is more expensive in terms of R-value per inch, but offers the maximum insulating value for its thickness.

Polystyrene Foam. This type of foam is made two ways. One process produces extruded board which is blue or pink. When cut, the many air bubbles inside are revealed. The other process produces expanded polystyrene (EPS) which consists of many tiny foam beads bonded together. As a result, EPS is commonly known as "beadboard." Extruded polystyrene is a slightly better insulator than EPS, and is more durable. Both types are well suited for insulating foundations and concrete slab floors.

Polyurethane/Isocyanurate. This efficient, versatile type of rigid plastic foam insulation is recognized by its white or yellowish color. Usually faced with foil, it is a good material for beefing up the R-value of insulated walls or ceilings. It is installed between the framing and wall finish, or between an existing wall and a new wall or ceiling finish.

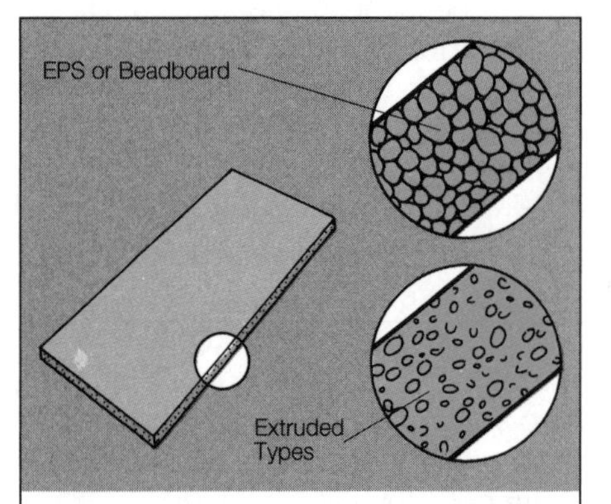

Polystyrene Foam. Expanded polystyrene insulation looks like many small beads pressed together, while extruded types look like soap bubbles.

Loose-Fill Insulation

When nooks and crannies cannot easily be insulated with batts, blankets or rigid insulation, loose-fill insulation answers the call. It comes in bags of pellets or fibers that are poured where needed or blown in by pneumatic equipment. Loose-fill can be used to add R-value to an underinsulated attic by pouring it over the top of existing insulation.

Caution: *Do not pour loose fill insulation into recessed light fixtures or other heat-producing equipment.*

Cellulose. Cellulose loose-fill insulation is made of recycled newspaper that has been treated with a fire-retardant chemical. It can be poured from the bag or blown in using special equipment. While the R-value of cellulose is slightly lower than fiberglass or mineral fiber, its greater density means that it is less susceptible to being blown out of place. The main disadvantage of cellulose is its tendency to attract and hold moisture. It is not a good material for houses that do not have solid moisture barriers.

Cellulose. Cellulose is purchased as loose-fill insulation in bags. It can be poured into hard-to-get places, such as enclosed walls and floors.

Perlite and Vermiculite Pellets. Perlite and vermiculite loose-fill insulation are noncombustible materials that are poured from a bag. Perlite is made from small pieces of aluminum silicate expanded by rapid heating. Vermiculite is made from mica. These materials are relatively low in R-value (2 to 2.5 per inch) and relatively high in cost. They are most useful for filling small cracks and voids that fluffier forms of loose-fill will not settle into, such as between blankets, around chimneys and framing and odd-sized cavities.

Fiberglass and Mineral Fiber. Since loose fibers composed of fiberglass or minerals must be installed with special pneumatic equipment; the job usually is best left to professionals. Blowing in loose-fill fiberglass or mineral fibers is a good, economical way to insulate uninsulated walls without having to remove the surface finish. It also is a quick way to add insulation to an attic, as compared to installing batts, blankets and rigid insulation. R-values vary with the density at which the material is installed. Loose fibers tend to blow away, so they must be protected from high winds in an open attic.

Radiant Barrier Materials

Highly reflective surfaces that block radiant heat energy cut much of the heat gained through roofs. Radiant barriers cut the cost of cooling in the summer and, to a lesser extent, the cost of heating in the winter. They are most effective and most money-saving for those living south of the Mason-Dixon Line (see page 10, "Targeting Insulation Levels").

Radiant barriers are available in the form of rigid sheets (foil-faced rigid foam or plywood sheathing) and rolls of flexible sheeting. Whatever its composition, in order to be effective

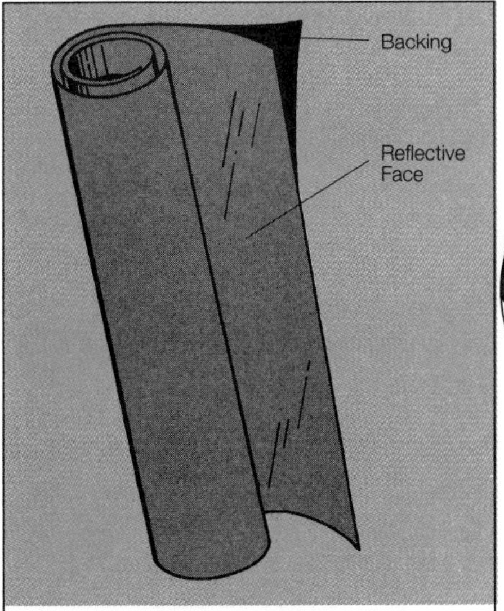

Radiant Barrier Materials. This insulation combines a reflective surface with a reinforced backing.

the barrier must be installed with an airspace separating the reflective surface from an adjacent material. Plywood sheathing that has one side of reflective surface is intended to be used on roofs in new construction (the reflective side faces inward). Low-surface emittance is a measure of how well a radiant barrier reflects radiant heat energy. Look for conformance to American Society for Testing and Materials (ASTM) C1158 on the label. Select a material with a backing strong enough to hold it together without tearing.

Pipe & Water Heater Insulation

Prevent energy from escaping from the water heater and hot water pipes by wrapping them with a specially made insulating product. Pipe sleeves are made of polyethylene foam for pipe diameters of 1/2 inch, 5/8 inch and 3/4 inch. Each package contains 36- or 48-inch-long pieces slit down their length for easy installation (simply press the slit over the pipe). Water heater blankets are packaged as kits that contain an insulated blanket encased in a plastic sleeve and self-stick tabs or tape. Simply wrap it around the water heater and secure it with the tabs or tape.

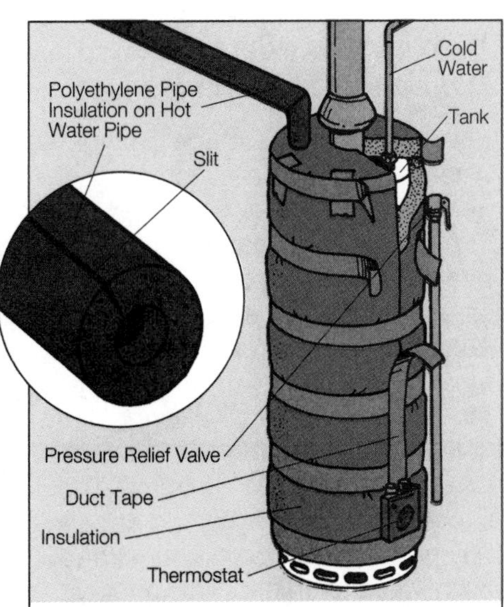

Pipe & Water Heater Insulation. This type of insulation is made of foam polyethylene.

Air & Vapor Barriers

To cut heat loss while gaining and preventing moisture damage it is necessary to reduce the passage of air and moisture through walls and ceilings. Selecting and applying the appropriate caulks, gaskets or sheet materials is an important part of the process.

Air Barrier House Wrap

This type of sheet material stops the bulk movement of air through walls. It is applied over the outer walls beneath the siding. Materials used for house wraps allow water vapor to escape to the outside, but at the same time keep out rain. Some materials, such as spun polyester, are able to do this because of their porosity. Other impermeable materials, such a polyethylene, are perforated with many tiny holes to make them permeable. Air barrier house wrap is available in rolls 9 feet wide for large wall surfaces, and in shorter widths for wrapping sill and head plates.

Air Barrier Tape

When the joints are sealed, plywood or waferboard sheathing performs double duty as an air barrier. Use a sealer made for this specific purpose. If you cannot find a sealer, use a moisture-proof tape that covers the joint and sticks to the substrate. Duct tape or contractor's tape (plastic tape designed specifically for use with air and vapor barrier sheet materials) work well. Although tape can be applied directly from the roll, a dispenser makes large jobs easier.

Vapor Barrier Materials

Unlike air barriers, vapor barriers block both air and water vapor. To do this they must be impermeable to moisture. When looking for a good vapor barrier material, check its permeance rating. To be acceptable as a vapor barrier, the permeance rating should be less than 1. Unpainted drywall has a permeance rating of 50 and is not considered a

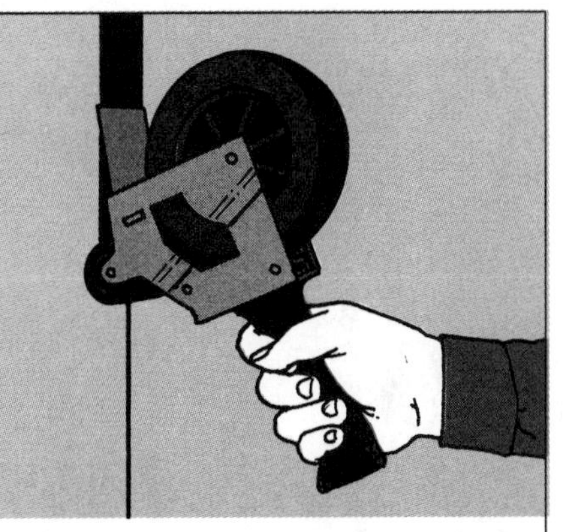
Air Barrier Tape. Joints in between panel sheathing can be sealed with duct tape or contractor's tape.

Air Barrier House Wraps. Spun polyester or perforated polyethylene house wrap comes in 3- and 9-ft.-long rolls.

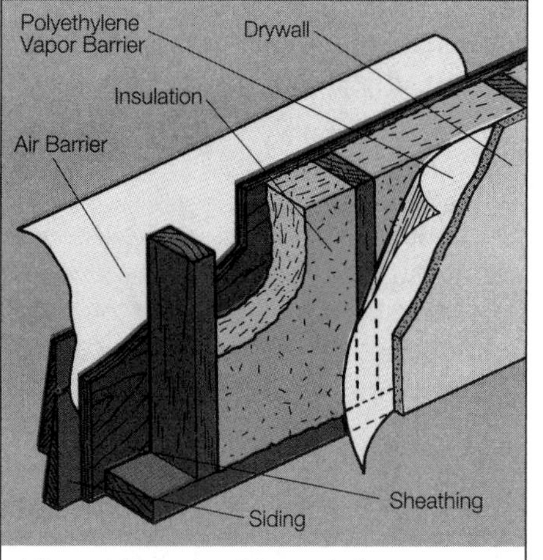

Vapor Barrier Materials. In a typical stud wall the air barrier is placed between the siding and sheathing. The vapor barrier is located on the warm side, just under the interior wall finish.

vapor barrier. The asphalt-backed kraft paper used to face blanket and batt insulation has a permeance rating of 1; it is a marginal vapor barrier. Two excellent vapor barriers are polyethylene sheet which has a permeance rating of .08 to .04, and aluminum foil which as a rating of 0.

Polyethylene Sheet. Polyethylene sheet is a material well suited to retarding vapor passage in a new floor, wall or ceiling. Polyethylene sheet is available in 8-foot-wide rolls that are 100 feet long and have thicknesses of 4,6 and 8 mils. A 6-mil thickness is preferred when working with walls or ceilings.

Weatherstripping & Sealing Materials

In order to create an effective barrier between the inside of the home and the outdoors, you must spend as much time plugging cracks as you spend installing insulation. As much as one third of an average home's heat loss in winter can be traced to leakage through cracks found around windows, doors, sills, electrical boxes and other gaps. Fortunately, there are a wide array of weatherstripping and sealing products available to seal up these openings.

Weatherstripping Window & Door Jambs

Weatherstripping is made with many different materials and comes in many shapes. Some are better suited to particular jobs than others. Choose packaged kits that contain rigid strips intended for average-sized doors or windows, or rolls of flexible material you cut yourself. Some types are applied simply by sticking them to the window or door while others have to be tacked to the frame. The success of weatherstripping depends on choosing the right product and installing it so that it seals snugly.

Foam Tape. Rolls of self-stick foam tape provide an economical way to seal light-duty jobs. After peeling off the backing strip, simply press it to the stationary part of a door or window. Though cheap and easy to apply, foam tape is the least durable of foam tapes. No matter how carefully installed, the foam deteriorates and loses its seal over time.

Rigid Strip. Rigid strips of vinyl, wood or aluminum are available in kits for weatherstripping single windows and average-sized doors. Usually the kits contain tacks for attachment.

Spring-Metal. This type of weatherstripping is made of copper strips, formed into wide V-shaped mounts that are installed inside the window track or door jamb. Cooper tacks hold them in place. Spring-metal weatherstripping is the most difficult to install, but when installed correctly, creates the best seal and lasts the longest.

V-Shaped Vinyl Strip. Like spring-metal, V-shaped vinyl strips attach to the inside of a window or door jamb. They employ a self-stick adhesive rather than tacks. To achieve a tight seal the material must be installed perfectly straight; curls result in air leaks.

Foam Tape. Peel-and-strip foam tape is the easiest weatherstripping to apply, but the least durable over time.

Rigid Strip. Flexible edges in rigid strips are available in kits of pre-cut lengths. They are tacked to non-moveable window or door parts.

Spring-Metal. These copper strips are installed with copper tacks on the inside of a window or door jamb.

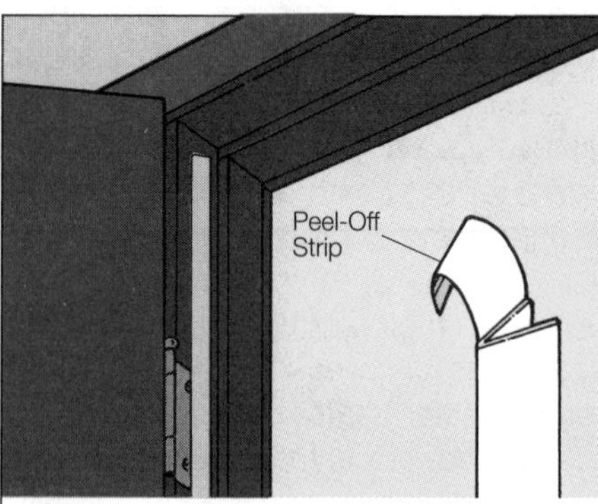

V-Shaped Vinyl Strip. This weatherstripping combines the tight spring seal of spring-metal with the installation ease of peel-and-strip.

Weatherstripping Door Bottoms

The bottom edge of a door is one of the biggest air leaks and the hardest to seal. Rigid strips of aluminum or wood that contain a flexible contact surface can be mounted to the door face or floor or to both.

Door Sweeps. Door sweeps mount to the bottom of the door and close off the crack between the door and threshold. The flexible edge may be vinyl, felt or brush. Most have oversized screw slots that allow the sweep to be adjusted up or down to obtain just the right contact with the threshold without preventing the door from closing completely. A sweep provides a quick, economical improvement for an exterior door that has a poorly sealed threshold. It also provides a good backup seal for doors that already have a good threshold.

Flexible Arch Thresholds. A good threshold is the first line of defense against drafts. One type contains a compressible vinyl arch mounted on a base of extruded aluminum or wood. The base can be screwed to a wood floor or secured by construction adhesive to a concrete floor.

Combination Thresholds. For doors subject to heavy rain, consider a two-part unit that combines a weather-stripped aluminum threshold with an overhanging shield. The top part sheds rain, while the threshold blocks drafts.

Garage Door Thresholds. A rubber gasket provides a snug fit between the bottom of a garage door and the floor.

Choosing Caulks & Sealants

Cracks, holes and gaps usually can be plugged with a gasket or a plasticized chemical compound. Picking the right one out of the dozens available can be confusing. Higher performance caulks naturally cost more, but because they last longer, the investment may be worthwhile. Because houses are built from materials that expand and contract with the weather, a caulk must retain the ability to flex, as cracks open and close.

Oil-based glazing caulks (glazing putty) do not remain flexible, so avoid them if possible. You'll also want to make sure the caulk will stick to the surface you are applying it to. (The surface may have to be primed with an exterior-grade paint primer first). Some caulks, such as silicones, cannot be painted unless specifically marked "paintable." Before buying, find out how messy the caulk is to work with and the types of solvent needed to clean it.

Tube and Cartridge Caulks. Most of the caulks listed in the chart come in cartridges for application with a caulking gun or squeeze tubes. For plugging cracks 1/4 inch wide or narrower, a bead of caulk suffices. Wider cracks require that a backer be installed before caulking to keep the material from penetrating too deeply. Backer rod made of foamed polyethylene can be used. It comes in rolls of 1/2-, 5/8-, and 3/4-inch diameter.

Foam in a Can. One of the most innovative and useful sealing products designed in the last decade is polyurethane foam sealant packaged in a pressure can. It can fill irregular, hard-to-reach cracks and gaps that cannot be plugged effectively with caulks or gaskets. The foam is very sticky and messy to work with because it expands in place, causing overflows until you get the hang of installing it. The excess can be trimmed with a knife. Always wear rubber gloves to protect your hands from inevitable spills. If a spill does take place, the only solvents that can be used to clean up are ketone-based, such as acetone and lacquer thinner. These solvents may irritate your skin.

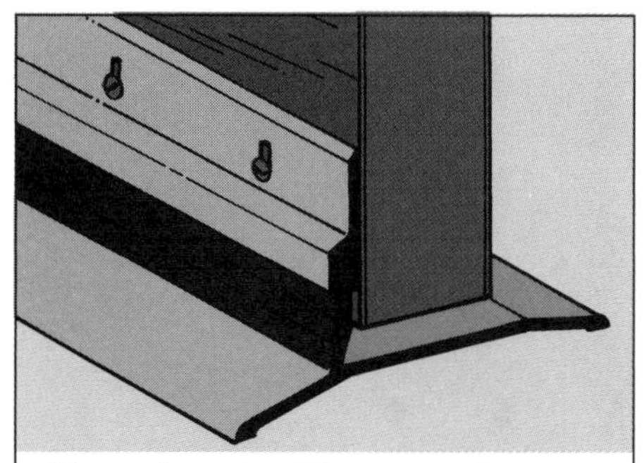

Door Sweeps. This type of weatherstripping consists of an extruded aluminum strip that holds a flexible vinyl strip. Brush and felt strips also are available.

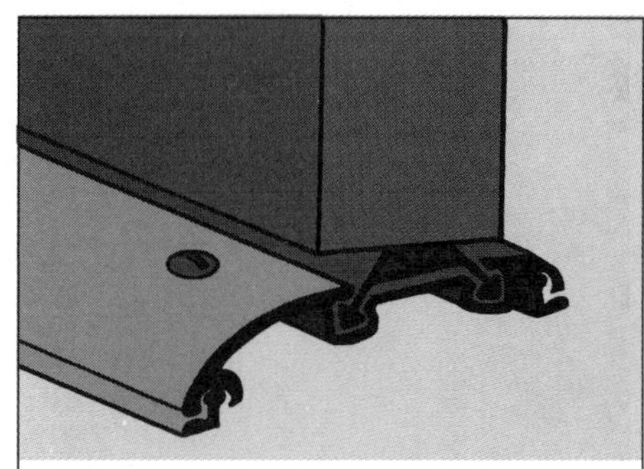

Flexible Arch Thresholds A flexible vinyl strip retained by a rigid aluminum base can be screwed to a wood floor or glued to a concrete floor.

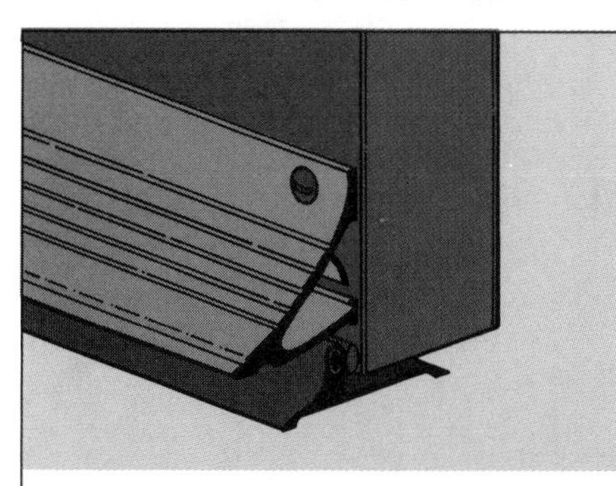

Combination Thresholds. An aluminum weatherstripping threshold below an overhanging shield blocks both water and air leaks.

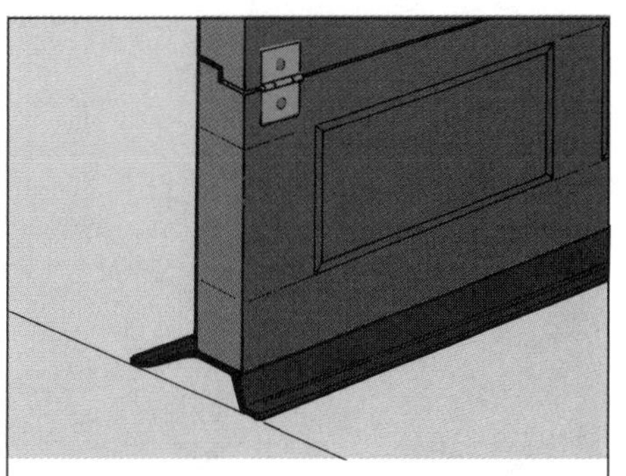

Garage Door Threshold. Flexible gaskets are available for specialized needs, such as the bottom edge of a garage door.

Choosing Caulks

CAULK	SPECIAL USES								Relative Cost (High, Moderate, Low)	Longevity in Years	Service Temperature Range (Degrees F)	Shrinkage Resistance (Excellent, Good, Fair)	Primer Needed on Porous Surfaces	Can be Painted	CLEAN-UP LIQUID		
	Interior Cracks	Exterior Cracks	Glass to Frame	Metal to Masonry	Concrete, Brick	Roofing, Flashing	Holes and Very Wide Gaps	Sealing Poly Vapor Barrier							Water or Soap and Water	Paint Thinner, Turpentine	Acetone
Acrylic Latex Caulk	●								M	3 to 10	-20° 180°	F	●	●	●		
Siliconized Acrylic Latex Caulk	●	●							H	15 to 25	-20° 180°	E	●	●	●		
Butyl Rubber Caulk		●		●	●			●	M	5 to 20	-20° 200	F		●		●	
Neoprene Caulk				●	●				H	15 to 20	-40° 250°	G		●[1]			
Polysulphide Caulk			●						H	10 to 20	-40° 250°	E	●	●[1]			
Silicone Caulk	●	●	●	●					H	20 to 50	-75° 450°	E	●	●[2]	●		
Asphaltic Caulk				●				●	L		40° 150°	F		●		●	
Polyurethane Caulk	●	●	●	●	●			●	M	20 to 30		M	●	●[1]		●	
Polyurethane Foam in Pressurized Container							●		H	20	40° 120°	F		●[3]			●

1. Can be painted only after curing (two weeks or longer).
2. Cannot be painted unless specifically indicated "paintable".
3. Paint any foam sealant exposed to sunlight.

Refer to the label on any product to adjust the above data and add any special conditions of use.

Windows & Doors

An old window that has single-sheet glazing loses as much heat through conduction as 12 square feet of wall that is insulated to R-11. In addition, most windows lose even more energy through cracks around the sash. Sit next to a leaky window on a cold winter night and you will know how uncomfortable it can be. Similarly, a door that is not weather-tight is a hidden source of energy loss. It does not make sense to insulate the walls beyond average levels without also improving the energy-saving performance of the windows and doors.

Weather-Tight Windows

The easiest way to seal cracks around the stationary parts of windows is to caulk them with a flexible caulk. Sealing cracks found around the movable sash requires weatherstripping. There are many types of weatherstripping from which to choose. Always follow manufacturer's instructions.

Double-Hung & Sliding Windows

Double-hung and sliding windows are responsible for the most insidious air leaks. The job of plugging the leaks often requires weatherstripping around jambs, heads, sills and meeting rails. The following instructions are for weatherstripping double-hung windows. The procedure is the same for sliding windows except you must treat them as if they were double-hung windows turned on their side.

1 Cut pieces of spring-metal weatherstripping 2 inches longer than the sash height. Raise the bottom sash and feed the top edge of the weatherstripping into the jamb. Extend the rest of the strip down to the sill and tack it to the jamb. Repeat the process in reverse for the top sash.

1 Weatherstripping Jambs. Spring-metal weatherstripping is the best choice for this application. It has a nice appearance when the window is open and is durable over time. Measure the height of the bottom sash and then use tin snips to cut pieces of spring-metal weatherstripping 2 inches longer than the sash height. Raise the bottom sash and insert the top edge of the weatherstripping from the bottom upward, feeding the extra two inches between the sash and jamb. Extend the rest of the strip down to the sill and tack it in place using the tacks provided. Repeat the process in reverse when installing stripping for the top sash.

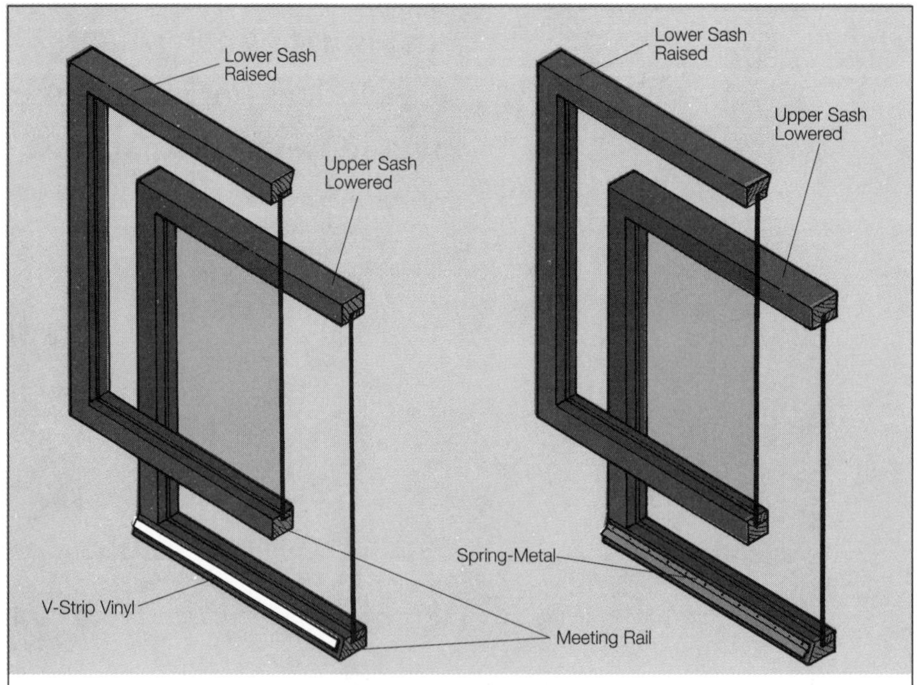

2 Use either V-strip vinyl (left) or more durable spring-metal (right) to weatherstrip the meeting rail.

2 Weatherstripping the Meeting Rail. Slide the bottom sash up and the upper sash down to allow access to the meeting rail of the upper sash. Apply V-strip vinyl or spring-metal weatherstripping to this rail. When using V-strip vinyl, attach it so that the V points upward and the open leg points downward. When using spring-metal, make sure the open end faces downward. Tack gently and have a helper hold a brick or other heavy object against the back of the rail to provide support for the glass.

3 Weatherstripping Head and Sill Rails. If you are using a window weatherstripping kit, tack the pieces designated for head and sill onto the top and bottom of those

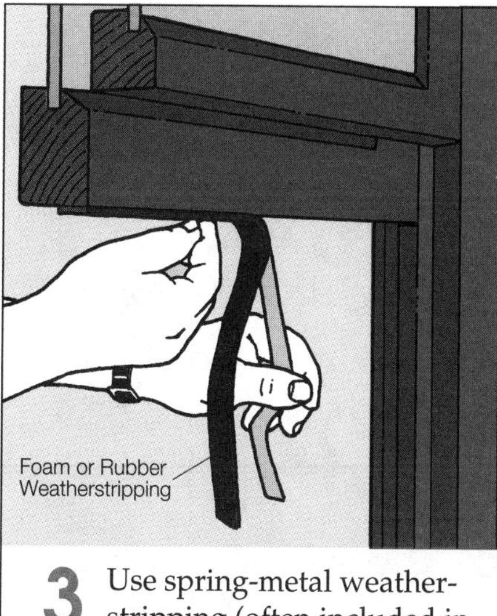

3 Use spring-metal weatherstripping (often included in kits) or a foam gasket to weatherstrip the top of the head rail and the bottom of the sill rail.

parts. However, if you are not using a kit, you can effectively seal the heads and sills with weatherstripping that compresses, such as V-shaped vinyl, spring-metal or foam backed with self-adhesive.

Wood Casements & Awnings

Tack spring-metal weatherstripping or apply self-adhering V-shaped vinyl weatherstripping around the inside of the jambs of wood casement and awning windows.

Metal Casements

Metal windows require the use of a special type of gasket that has grooves. Slip the groove over the jamb, all around the frame.

Weather-Tight Doors

When it comes to insulation, energy-efficient doors are equally as important as well-insulated walls and weather-tight windows.

Insulating a Door

Before applying weatherstripping to a door, make sure the door opens and closes without binding. If necessary, adjust the hinges and rasp off edges that bind the frame. As with windows, weatherstripping for doors is available in kits that contain enough material for an average-sized exterior door. The weatherstripping must be durable enough to remain flexible and maintain a tight seal over a long period of time, especially if the door is opened and closed frequently. It probably is best to pay a little more for quality weatherstripping, such as those comprised of flexible material encased in rigid vinyl, wood or aluminum, than to buy cheaper material cut from a roll. In addition, rigid strip weatherstripping is easier to install. The following instructions show how to apply rigid aluminum strip weatherstripping that contains a vinyl bulb.

Wood Casements & Awnings. Attach spring-metal or vinyl V-shaped weatherstripping around the inside of the jambs of casement and awning windows.

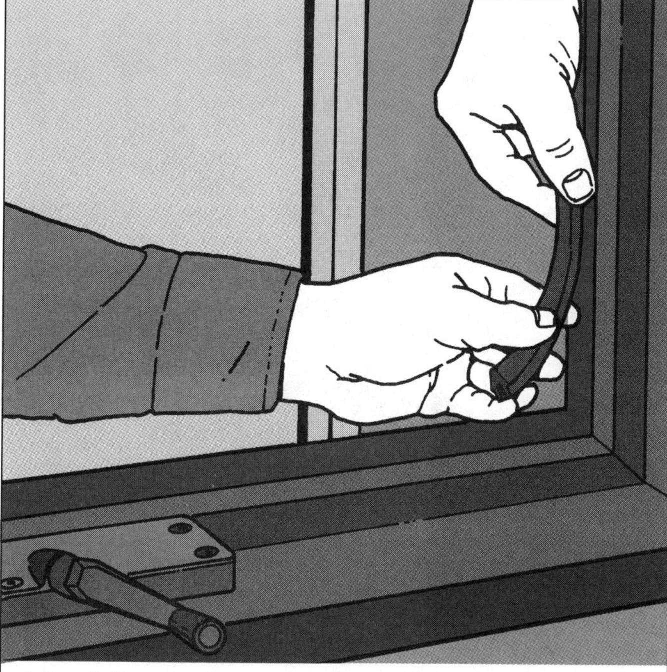

Metal Casements. Slip a deeply grooved gasket over the jamb of metal casements and awnings.

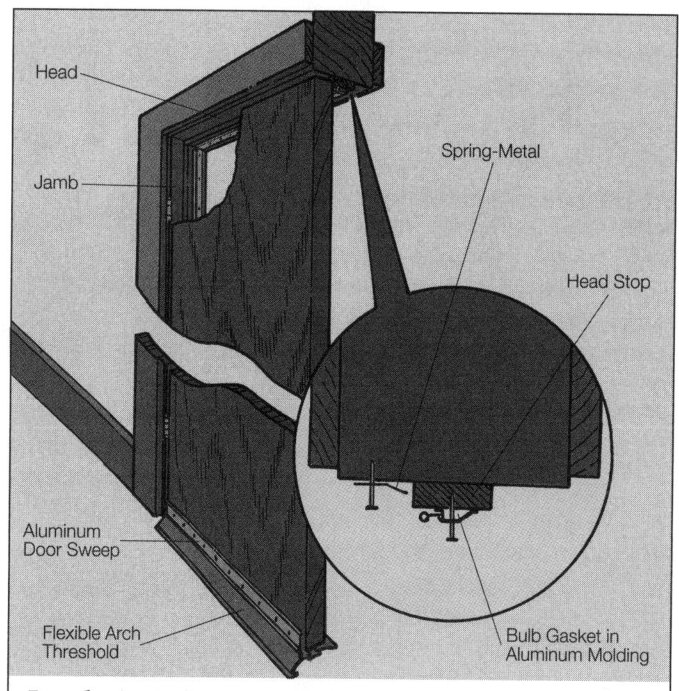

Insulating a Door. Use a spring-type weatherstripping inside door jambs or a compressible gasket on the stops. Seal door bottoms with thresholds and sweeps.

1 Measure the head strip and use a hacksaw to cut it to length. A block that is nailed to a board creates a cutting guide.

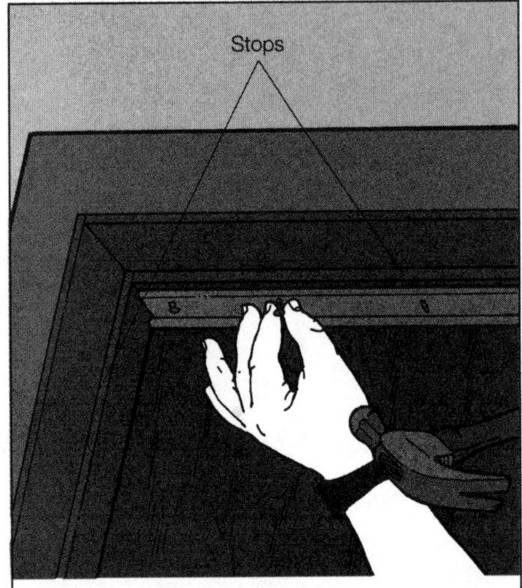

2 Driving the tacks only partway, tack each piece in place with the door closed.

3 After attaching strips to jamb stops adjust the fit by sliding paper between the door and flexible strip.

1 Measuring and Cutting the Head Strip. Measure the width across the top of the door opening from the inside of the stops. Then use a hacksaw to cut a length of weatherstripping to this dimension. Nail a piece of scrap wood to a board to serve as a cutting guide.

2 Attaching the Head Strip. Close the door and hold the head strip snugly, but not tightly, against the door. Use the tacks provided to secure it to the stop, driving the tacks deep enough to hold the piece in place (do not drive them tightly at this point).

3 Installing Jamb Strips. Measure the height of the jamb stops from the head strip to the floor. Cut two lengths of weatherstripping to this measurement. With the door closed, attach the strips to the jamb stops in the same manner the head strip was attached. When both strips are loosely secured, slide a piece of paper between the door and the flexible edge of the weatherstripping. Adjust the position of the strips until the paper barely slides. If the paper slides easily, the weatherstripping is not snug enough. If the paper does not slide at all, the weatherstripping is too tight. Drive in the tacks all the way. Then adjust the head strip and drive in the tacks completely.

Sealing the Door Bottom

If the door presently has a leaky threshold the easiest way to stop air from blasting through the crack at the bottom is to attach a door sweep to the bottom edge of the in-swinging side. A better seal is made by replacing the threshold with one that contains a compressible gasket, or one of the other airtight thresholds available at home center stores.

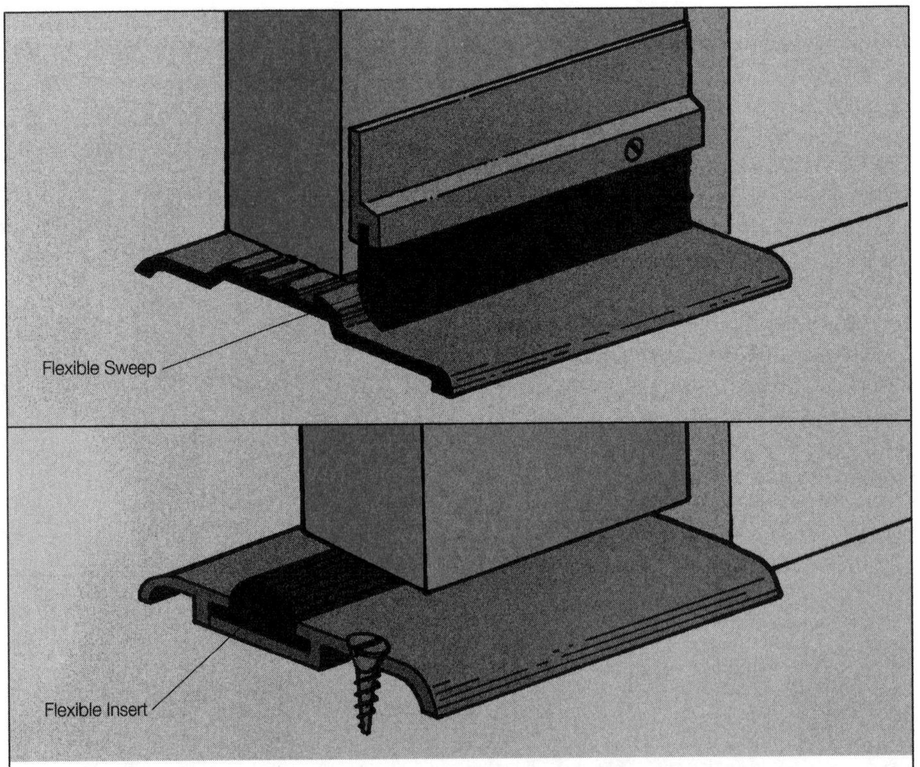

Sealing the Door Bottom. The gap between a door bottom and standard threshold is plugged by attaching a door sweep to the inside of the door (top). An even better solution is to replace the threshold entirely with one that contains a flexible strip (bottom).

Attaching a Door Sweep

If necessary, use a hacksaw to cut the sweep to the exact length between door stops. Place it on the bottom edge of the side of the door that swings toward you when you open it. Done this way, the sweep fits snugly to the threshold when the door is closed. Screw the sweep to the door through the slotted holes. Before tightening the screws, adjust the fit so that the door closes easily.

Installing a Threshold

1 Removing the Old Threshold. First try to pry up the existing threshold using a pry bar. If the threshold does not yield it has to be cut. Protect the floor finish by taping a piece of cardboard or thin plywood in the area to be cut. Then use a backsaw to make the cuts and pry up each end.

2 Installing the New Threshold. Refer to the manufacturer's instructions that come with the threshold. If the threshold has a flexible arch insert, set it in position over a bead of caulk. Lift up the insert and screw the threshold to the floor. Replace the insert and seal the outer edge of the threshold with a bead of caulk.

Attaching a Door Sweep. Cut the sweep to fit between door stops. Place it on the inside bottom edge of the door so that it contacts the threshold when the door is shut. Screw it to the door through the slotted holes. Adjust the fit so that the door closes easily before tightening the screws.

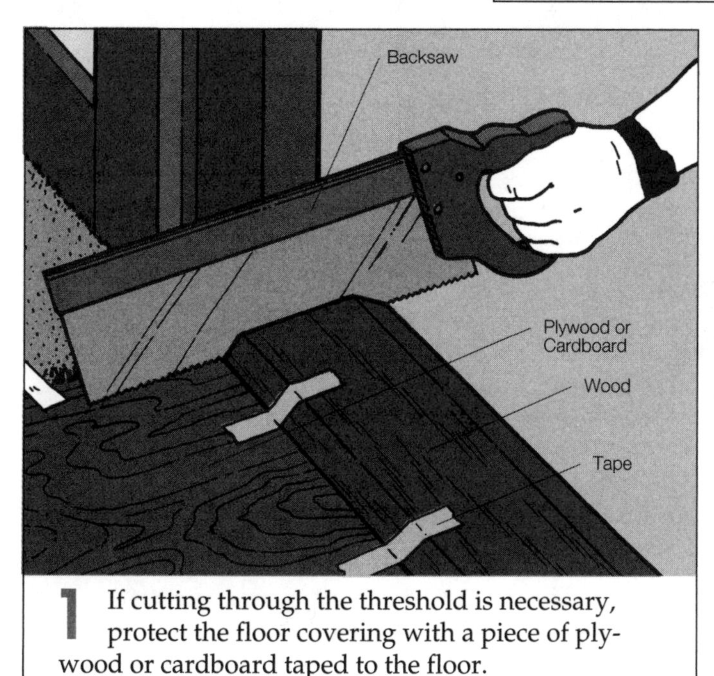

1 If cutting through the threshold is necessary, protect the floor covering with a piece of plywood or cardboard taped to the floor.

2 Cut the threshold to fit between jambs and set it into a bead of caulk. Lift up the insert and drive in the screws. Replace the insert and apply a bead of caulk to the outer edge.

Weatherstripping a Garage Door

Because garage door sections do not close snugly against the jambs like swinging doors do, sealing garage doors tightly is next to impossible. Still, you can cut the river of incoming cold air down to a stream by weather-stripping the head, sides and sill. Most likely, everything you need to do the job will be found in two separate kits: one that contains a flexible vinyl or rubber flange to go around the top and sides, and another that contains a rubber gasket to fit onto the door bottom.

1 Weatherstripping the Top. Work from the outside with the door closed. Measure the clear distance between jambs and cut a piece of weather stripping to this length. Press it snugly into the door face, and nail it to the head with the nails provided in the kit.

2 Attaching the Sides. Measure side pieces so that they extend from the flat part of the head weather-stripping to the floor. Then cut the pieces to this length. Next, trim a triangular piece off the flange of the top of each piece, fitting it to the flange of the head piece. Press each piece snugly into the door and drive the nails.

3 Attaching the Bottom Seal. Use scissors or a utility knife to cut the gasket to the full width of the door. Open the door partway and prop it up with a piece of scrap lumber. With the thickened edge of the channel on the outside, fit the gasket onto the door bottom and nail it in place with roofing nails.

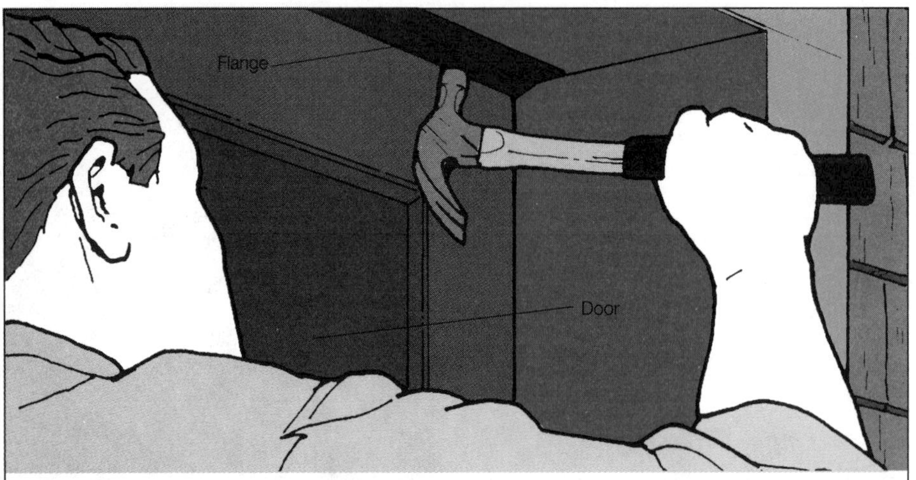

1 Cut a piece of weatherstripping to fit between the jambs at the head and press it snugly into the door face. Use the nails provided to attach it to the head.

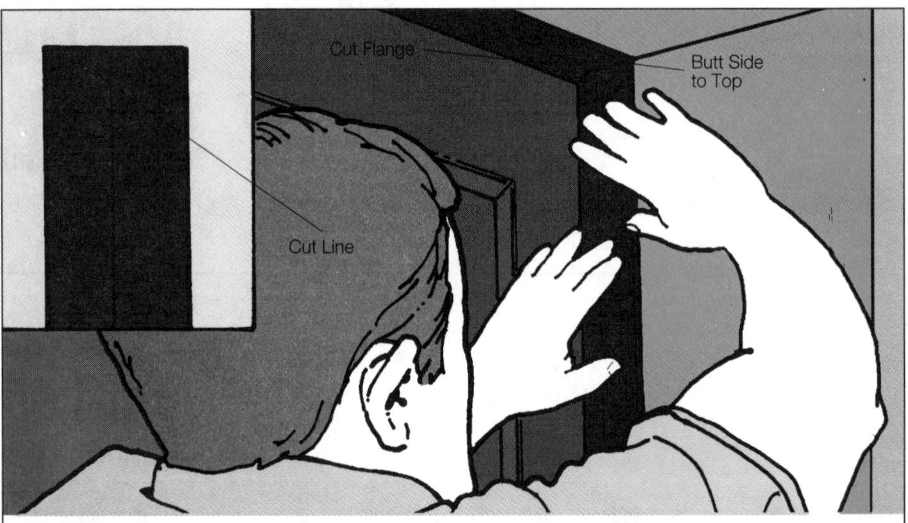

2 Cut side pieces to fit between the floor and the flat part of the head piece. Trim a triangular piece off the flange of the top of each piece, fitting it to the flange of the head piece. Press the pieces into place and drive the nails.

3 Cut the sill gasket to the full width of the door. Prop open the door partway and, with the thicker edge of the channel on the outside, fit the gasket onto the door bottom. Use roofing nails to nail it in place.

INSULATING WALLS

There are many ways, both major and minor, to make your walls more energy efficient. If you are renovating a house or a room, you might go as far as removing the walls and adding insulation batts. A lot can be accomplished, however, simply by sealing up all the nooks and crannies you can find.

Stopping Air Leaks

Cutting down the flow of air that passes unchecked through a home's weather protection is the best and cheapest way to trim heating bills. It also plays a major part in improving indoor comfort and solving moisture problems.

Sealing Leaks from the Inside

The easiest way to detect the source of an air leak inside the home is to wait for a cold, windy day, then hold your hand up to every object in the outer wall including windows, doors and electrical outlets. A tissue can help detect air movement. Hold it up to a suspected crack and see if it flutters. Smoke from a stick of incense can be helpful in a similar manner.

Once the leaks have been tracked down they must be plugged with the appropriate material. Attach weather-stripping to doors and

Caulking Leaks on the Outside

Look for obvious cracks around windows and doors, gaps between the siding and the foundation, and holes where wires and conduits enter the house. Fill narrow cracks with flexible caulk that is rated for exterior use (see page 20, "Choosing Caulks"). Use foam sealant to seal up larger gaps.

Caution: *Never caulk the horizontal joints in lap siding. Doing so cuts off the escape route of moisture and leads to the deterioration of paint and siding.*

1. ***Removing Old Caulk.*** Use a pointed tool such as a chisel or putty knife to remove the existing caulk before applying fresh caulk. Blow out dust that collects at the bottom.

2. ***Inserting Backing Rod.*** Before caulking, insert a polyethylene foam backer rod into joints that are wider than 3/8 inch.

3. ***Applying Caulk with a Caulking Gun.*** Squeeze an even bead of caulk into the cleaned joint.

4. ***Tooling the Joint.*** Run your finger over the caulk to imbed it into the crack, and at the same time, create a concave profile. Use water or a solvent (indicated on the label) to remove excess caulk from the sides.

windows (see previous chapter) and seal other openings with a suitable caulk. Do not rely on fiberglass insulation when it comes to stuffing the cracks. Fiberglass is a good barrier against conduction, but a poor one against convection.

Sealing Electrical Boxes at the Wall Face. The easiest way to seal off the air pathway through an electrical box is to remove the cover plate, place a foam insulating pad over the switch or receptacle and replace the cover. Foam pads are made with holes to match the profile of the device within the box.

Sealing Leaks from the Inside. Hold a loose tissue or lighted incense stick near suspected leaks.

Using a Caulking Gun

With the help of a caulking gun most cracks can be sealed easily. Nip off the end of a cartridge of caulk and stick a long nail or wire into it, puncturing the seal. Then pull back the plunger of the caulking gun and insert the cartridge. Hold the gun at a 45-degree angle and squeeze a bead of caulk into the crack as you move the gun along its length. Release pressure on the trigger just before reaching the end. Unless continuing immediately to another crack, pull out the plunger to prevent caulk from running out the tip.

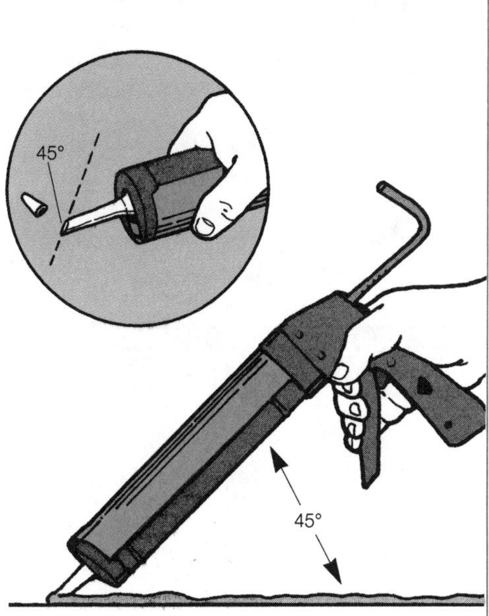

Caulking Electrical Boxes. The surest way to completely seal off an electrical box is to caulk the openings inside the box. Remove the receptacle or switch and inject silicone or polyurethane caulk around wire penetrations inside the box and between the box and the drywall or plaster.

Caution: *Always shut off power to the circuit at the service panel before removing a switch or outlet cover.*

Sealing Around Band Joists. The band joists, also known as box joists or stringer and header joists, run around the perimeter of the house

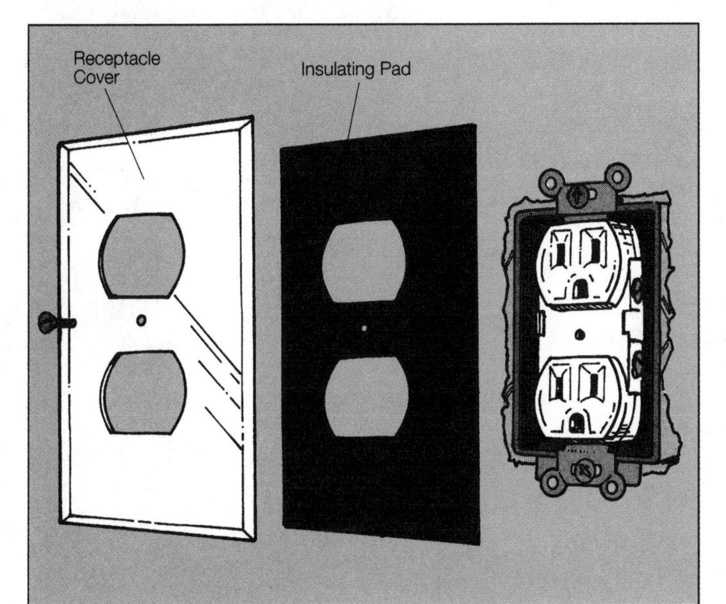

Sealing Electrical Boxes at the Wall Face. Plug up the airstream that flows into and out of boxes by placing a foam insulating pad over the switch or receptacle.

Caulking Electrical Boxes. Shut off the power to the circuit. Then remove the receptacle or switch and inject silicone or polyurethane caulk around all wire penetrations inside the box, and between the box and the drywall or plaster.

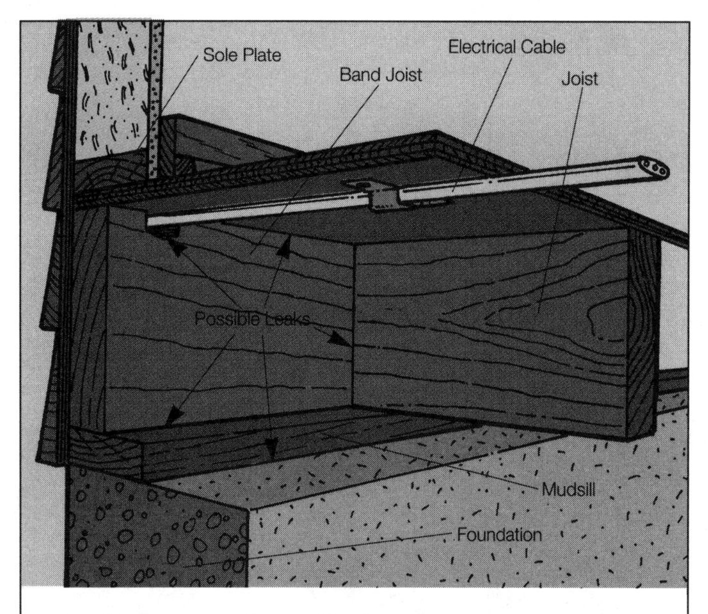

Sealing Around Band Joists. Seek out possible air leaks around the band joist and caulk them with a flexible caulk or inject them with foam from a can.

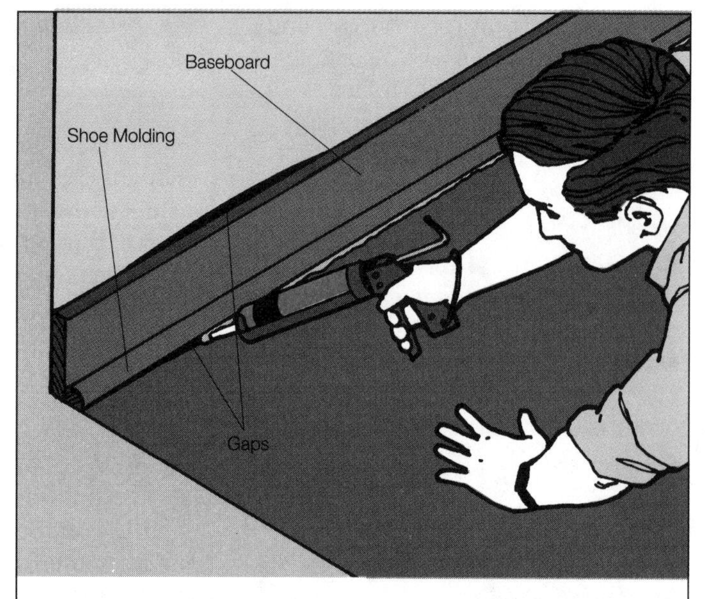

Caulking Floor-to-Wall Joints. Use a flexible caulk to seal gaps between the floor and base shoe, and between the baseboard and the wall.

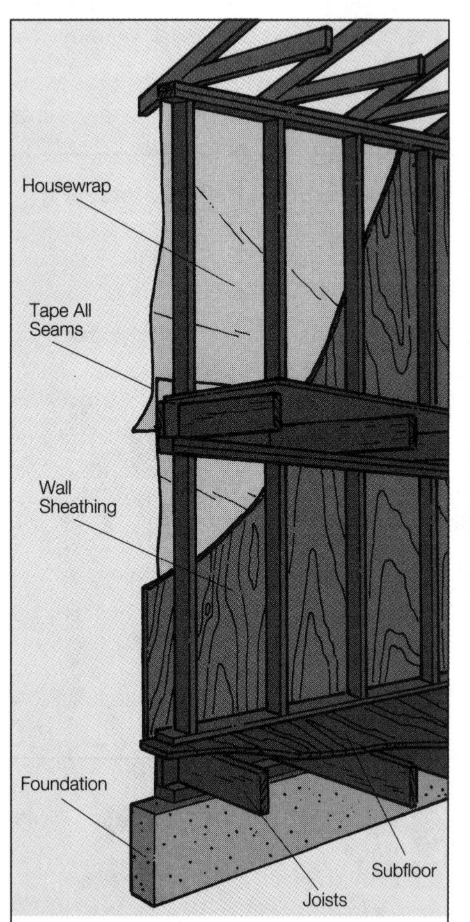

Installing Air Barrier House Wrap. Staple house wrap to the sheathing, covering all surfaces completely. Then use contractor's tape to tape the seams.

under the floor. Air enters through many points in and around a band joist. Use a flexible caulk or pressure foam from a can to plug leaks.

Caulking Floor-to-Wall Joints. Hold your hand in front of suspected leaks between the baseboard and wall, and between the floor and base shoe. Renail loose parts, if necessary, and caulk the gaps with a flexible caulk, such as polyurethane.

Creating an Air Barrier

Replacing siding and building new walls are two projects that provide opportunities to install insulation.

Installing Air Barrier House Wrap. Staple the house wrap to the sheathing. Cover all surfaces including window openings completely. Then slit and remove the parts that cover the windows and tape all seams with contractor's tape (a 2-inch-wide plastic tape made for this purpose).

Taping Joints with Air Barrier Tape. An alternative to using house wrap is to tape the joints between sheathing panels, and sheathing and door and window penetrations. Use air barrier tape designed for this purpose. Apply the tape by hand or, for larger jobs, use a dispenser.

Taping Joints with Air Barrier Tape. An air barrier is created by using house wrap tape to tape all joints in the sheathing and around windows.

Installing Thermal Insulation

The walls found in homes located in every climate require some insulation, as indicated on the map on page 10, "Targeting Insulation Levels." Determine the amount of insulation the outer walls contain by removing the cover plates from a few electrical outlets on different walls. With the aid of a flashlight, look into the wall at the side of the box. Be sure to turn off the electrical circuit before removing the plate.

There are several ways to insulate an uninsulated wall and to increase its R-value. Choosing the right way depends on many things, such as how much insulation already is in place, how much you want, and whether the interior finish will be removed for other reasons.

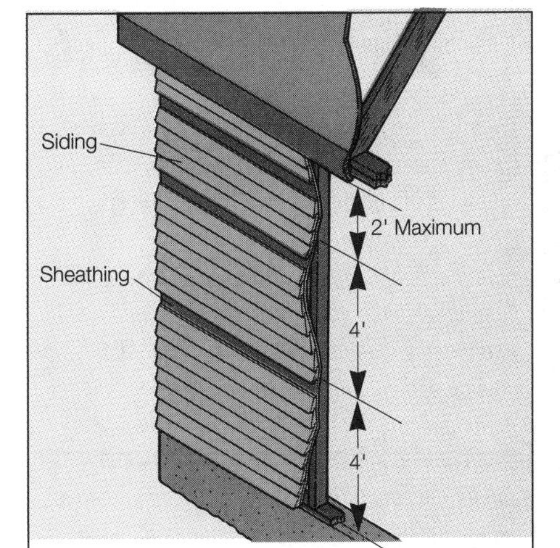

1 Starting 4 ft. up from the floor line, use a pry bar and hammer to remove strips of siding at 4-ft. intervals.

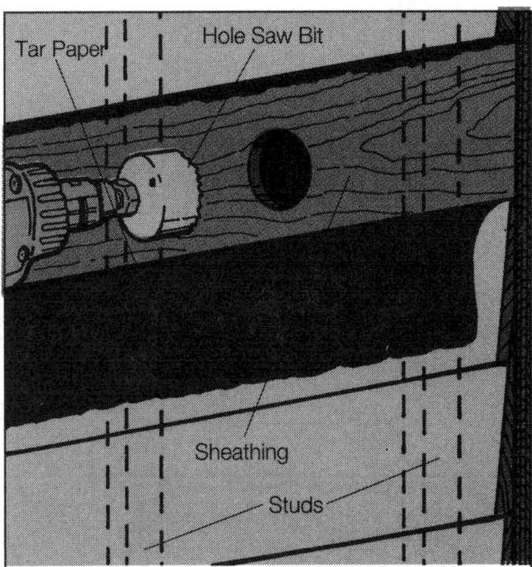

2 Cut tar paper along the top edge of one strip and pull it down to expose the sheathing. Drill through sheathing into each bay.

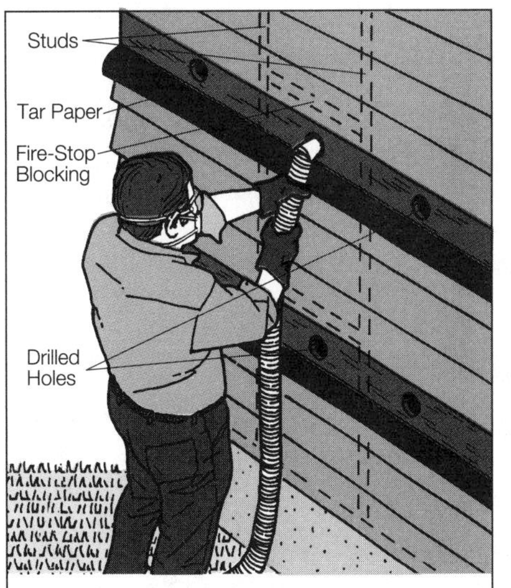

3 Starting at the bottom, insert the blower nozzle into each hole and fill each bay with insulation.

4 Plug the holes with a 3-in.-diameter cork or plastic plug and replace the siding.

Blown Insulation

Cellulose and fiberglass loose-fill insulation are blown into an attic or existing wall cavities by way of pneumatic equipment. This method minimizes damage done to the wall finish and is a fast, effective way to fill the stud cavities of houses built with regularly-spaced studs. It is less effective for post and beam houses which have irregular framing that results in odd-shaped spaces (such as triangular spaces between the braces) that are hard to fill.

The installation of blown insulation requires special equipment and expertise, so hiring an insulating contractor may be the best bet. However, those who have experience working with this material may still opt to do it themselves. Some building supply centers rent out the necessary equipment.

1 Removing Siding. Use a pry bar to remove strips of siding on the wall about 4 feet apart, starting 4 feet from the floor line. Use a chisel as a wedge along the length of the siding boards. After the first strip is removed, drill a hole through the sheathing and stick a length of coat hanger inside to make sure that the strip is below the horizontal fire-stop. The topmost strip can be no further than 2 feet to the top of the wall (limited by the distance the insulation can be blown upward).

2 Drilling Holes. If tar paper was installed below the siding, cut it along the top edge of a strip exposed by the removed siding. Pull down the flap and note the pattern of nails on the sheathing. Most likely, the stud placement will be evident. Shut off the power that goes to circuits found in the walls that require drilling. At each strip of removed siding, use a 3-inch-diameter hole saw bit to drill a hole through the sheathing into each bay of studs.

3 Filling the Bays. Refer to the manufacturer's instructions that come with the blowing machinery for loading the hopper and operating the equipment. Starting at the bottom, insert the nozzle into each hole and fill each bay with insulation.

4 Replacing the Siding. Plug the holes with a 3-inch-diameter cork or plastic plug; then replace the siding.

Adding Insulation From the Inside

There are many good reasons to add insulation to the inside of a wall rather than the outside. Except for blown insulation, insulating the outside entails the cost of new siding. Working outside also puts you at the whim of the weather. In addition, reaching the top story of a two- or three-story house requires scaffolding. Finally, in most cases an outside job has to be done all at once, whereas inside work can be finished room by room.

The following step-by-step procedures describe two ways to add insulation to the room side of an existing wall. One requires rigid foam, the other uses blankets or batts. Vary the method to suit the job at hand. Things to consider include whether or not the wall finish has to be removed, and the amount of floor space lost to the thickened wall.

Blowing from the Inside

Insulation can be blown into walls from the inside, but most people try to avoid doing so because of the 3-inch holes that have to be to patched. With so many patches, this method may not be easier than gutting the wall finish and filling the cavities with blanket or batt insulation. After drilling holes at the top of each stud bay insert the hose (directing it downward) as far as possible. As the cavity fills, the insulation stops coming through the hose. Move the hose from side to side, slowly pulling it out of the wall as the wall fills. Pull out the hose completely only when the insulation has stopped flowing.

Rigid Foam Between Strapping

By applying 1½-inch rigid foam sheets set between 2x3 or 2x4 wood strapping, up to R-12 (depending on the type of foam used) can be added to the inside face of a wall. The strapping is applied directly over existing plaster or drywall, or to bare framing. Either way, it allows new wiring to be run between the rigid foam and the new wall finish.

If the aim is to add R-value and achieve a vapor barrier, leave the old wall finish in place. Before applying the strapping, make sure to remove the casings around doors and windows, baseboards, electrical cover plates, and other devices attached to the wall. After the wall has been strapped and insulated, the new finish requires each of these items to be reinstalled on the new wall face.

1 Locating the Studs. Locate the position of two studs by slowly passing an electronic stud finder over the wall surface, or by drilling a line of small-diameter holes across part of the width of the wall at mid-height. When the center of two consecutive studs has been found, measure the space between centers and use it to mark each stud across the wall. Drive a test nail at these locations to make sure a stud actually exists at each location. Use a carpenter's level positioned vertically to draw a line from the floor to the ceiling at each stud center.

2 Marking the Wall for Strapping. Starting at the floor make marks up each corner of the wall every 24 inches. These marks represent the centerline of the strapping. Next, make another mark above each of the marks at a distance equal to one half of the width of the strapping (for example, 1¼ inches for 2x3 strapping). Have a helper assist in snapping a chalkline between the marks to indicate the position of the top of each piece of strapping.

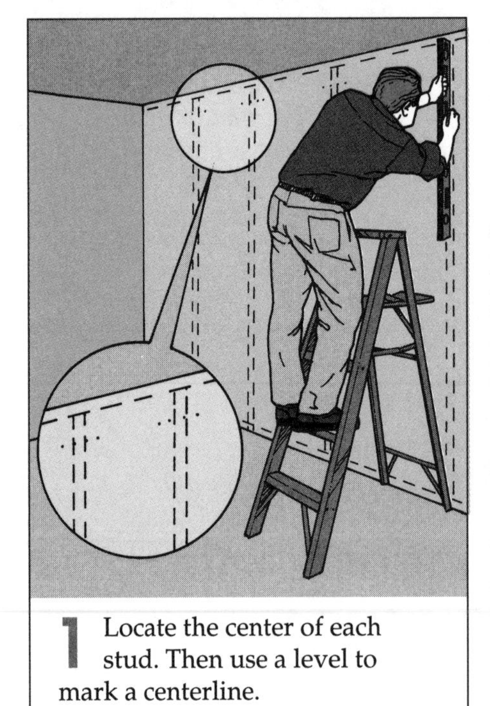

1 Locate the center of each stud. Then use a level to mark a centerline.

2 Lay out horizontal lines 24 in. on center.

3 Nail or screw each piece of strapping through the plaster or drywall and into the studs.

3 Attaching Strapping. Place each piece of strapping with its flat side against the wall. When working with drywall, use 20d galvanized nails to nail the strapping in place. For plaster walls, use 3½-inch galvanized deck screws to avoid breaking the plaster away from the lath with hammer blows. This is easier when the strapping has been predrilled.

4 Meeting Windows and Doors. Remove all casing and finish trim that protrudes from windows and doors. If the windows have protruding stools, usually it is easier to remove them and replace them later with wider stools than it is to try to cut them flush and add extensions. Install strapping around the openings. Cover the old casing areas but stop at the outside of the window and door jambs so they can be extended later. If the entire window is being replaced, install the strapping in line with the rough framing found around the window.

5 Framing for Switches and Receptacles. After shutting off the power to the circuits, detach the box and pull it a few inches out of the wall. Add a vertical cleat (made of the same stock as the strapping) to support the box. Attach the box to the cleat. If there is not enough wire in the box, try repositioning the box slightly to gain some slack.

Note: *Do not leave the old box in place and make a connection to the new box. Concealing electrical boxes is a fire hazard that violates the National Electrical Code.*

6 Running New Wiring. If you want to run new wiring you can use the spaces between strapping to do so as long as the cable is held 1¼ inches behind the face of the new wall surface (as required by the National Electrical Code). If it is not possible to hold back the cable 1¼ inches, nail a metal protection plate over the strapping through which cable passes. Use electrical staples to attach the wire to the back edges of the strapping. Cut a groove into the corner of the foam to provide a space for the cable.

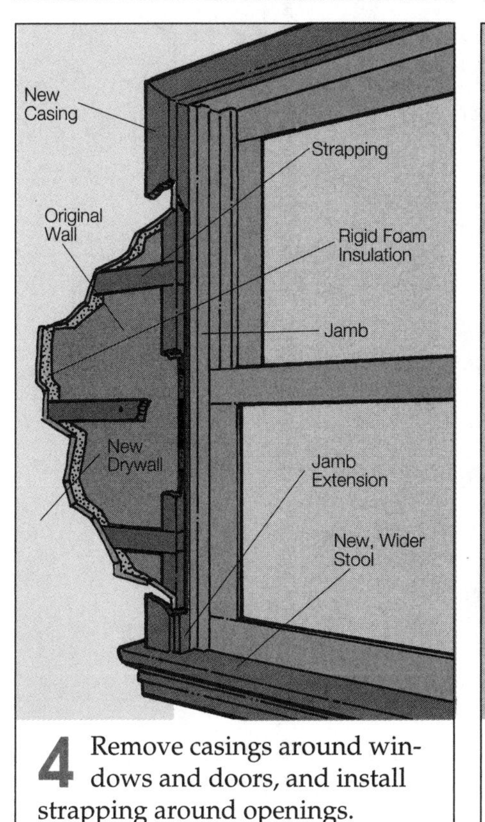

4 Remove casings around windows and doors, and install strapping around openings.

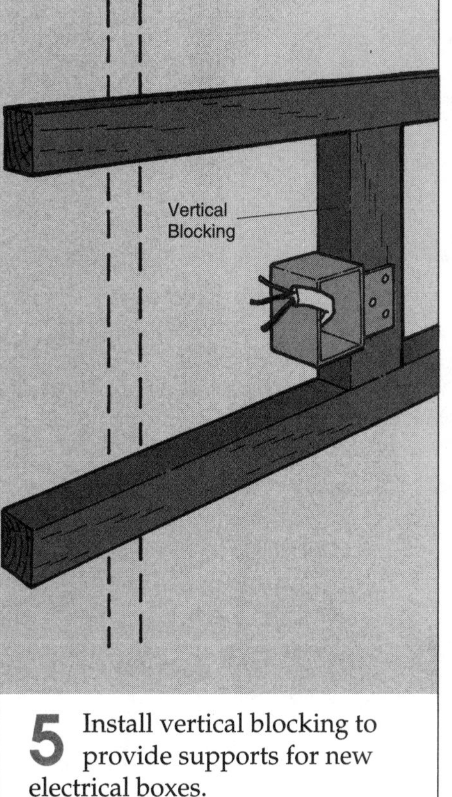

5 Install vertical blocking to provide supports for new electrical boxes.

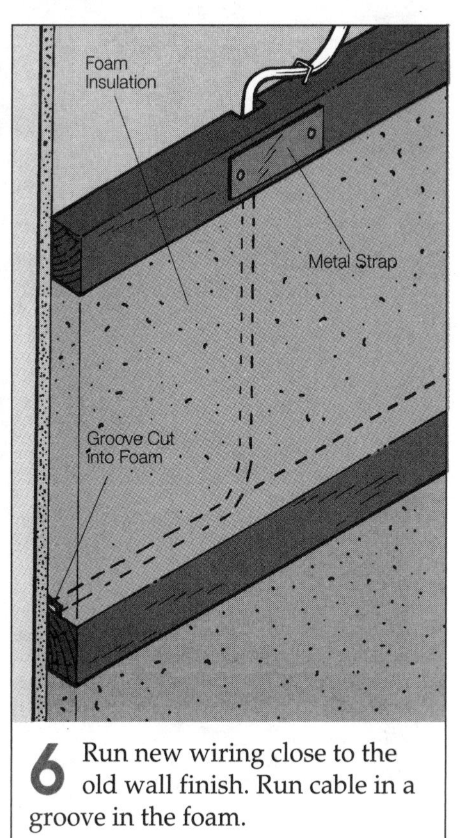

6 Run new wiring close to the old wall finish. Run cable in a groove in the foam.

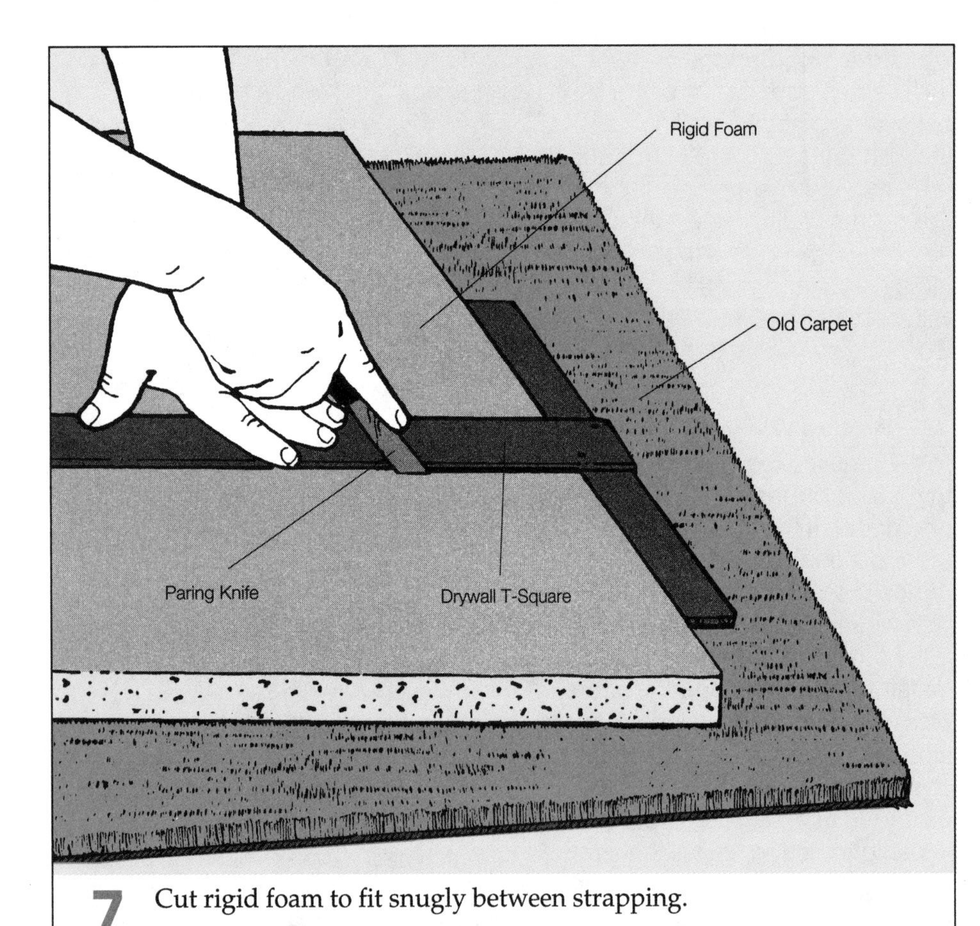

7 Cut rigid foam to fit snugly between strapping.

7 Cutting the Foam. Lay a sheet of rigid foam insulation over a cutting surface such as plywood or carpet. Measure and mark pieces to fit between the strapping (for example, 22½ inches for 2x2 strapping). Use a drywall T-square or a straightedge as a guide to cut through the foam. Then make two or three passes with a paring knife to cleanly cut through the foam and facing material.

8 Inserting the Foam. To hold the foam panels in place until the next step, place tabs of duct tape at the joints at corners and midpoints. Do not rely on friction to provide a tight fit.

9 Completing the Vapor Barrier. When using unfaced foam apply a 6-mil polyethylene vapor barrier over the foam and strapping as shown in "Rigid Foam over Gutted Walls," page 35. For foil-faced foam, use duct tape or contractor's tape to tape all joints between foam and strapping and window or door frames. Make two or more passes to

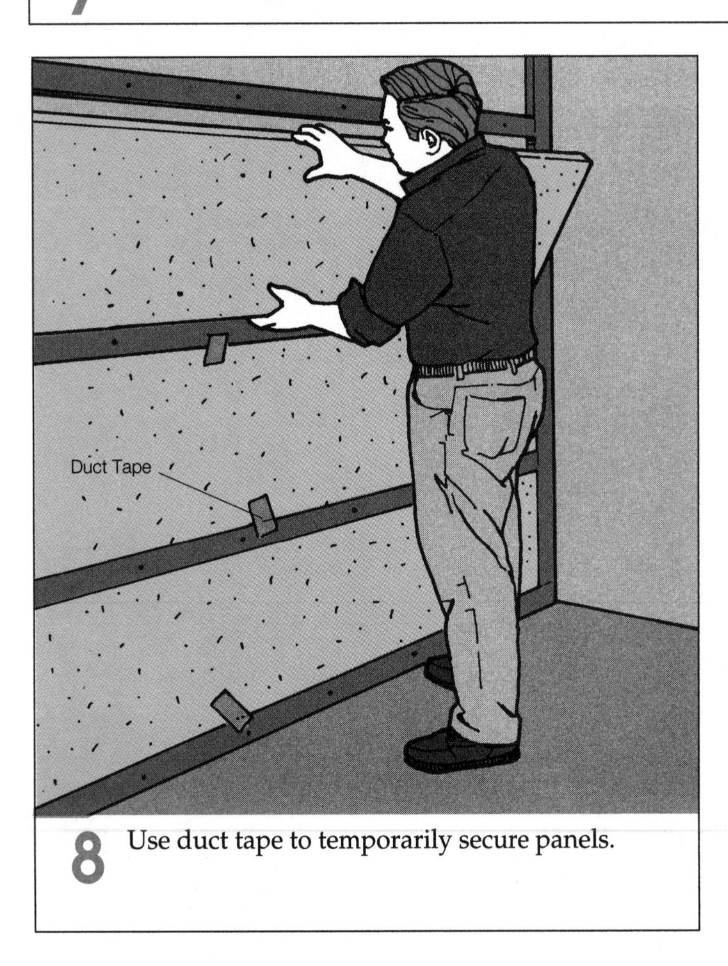

8 Use duct tape to temporarily secure panels.

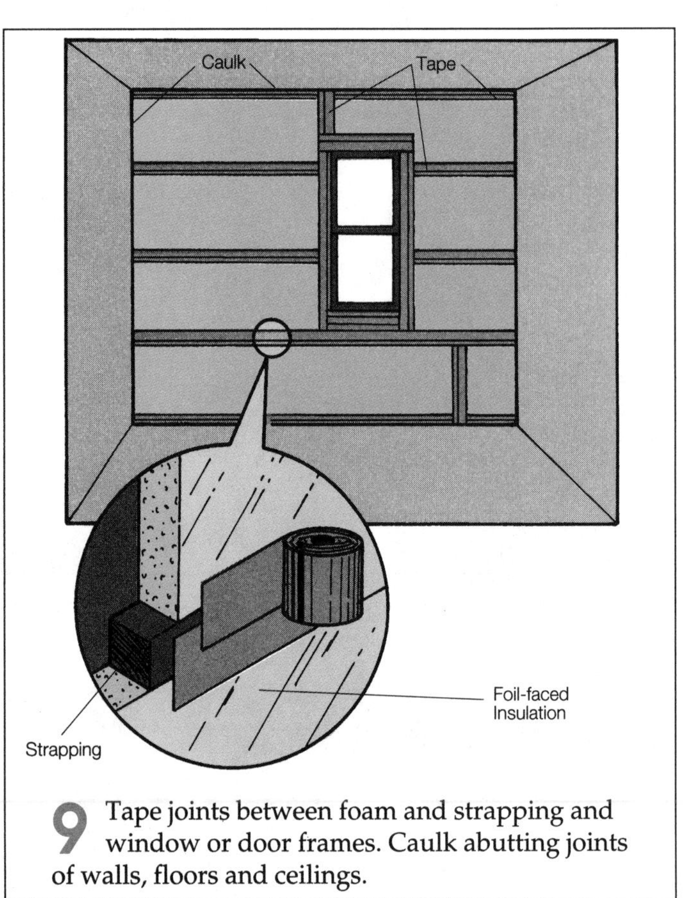

9 Tape joints between foam and strapping and window or door frames. Caulk abutting joints of walls, floors and ceilings.

Rigid Foam over Gutted Walls

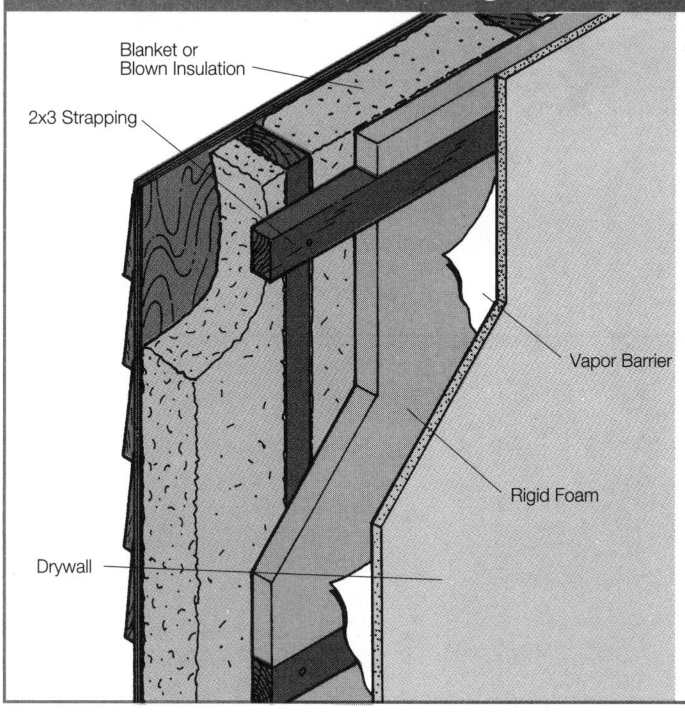

The existing wall finish may have to be gutted in order to insulate between the cavities, modify plumbing or change windows. In most cases just leaving or adding blanket or blown insulation between the bays provides all the necessary insulation. However, the width of the studs limits the thickness of insulation that can be added, so in cold climates it may be worthwhile to add rigid foam over the studs. This approach can add up to R-9 to the cavity insulation, while encroaching less than 2 inches into the room.

Those who decide to add a layer of rigid foam may be able to apply it directly to the framing before installing the new wall finish. However, drywall is easier to secure to strapping than to 1½ inches of foam, and strapping provides regularized supports if the existing framing is not spaced at 16- or 24-inch intervals.

After gutting off the finish, insulate the cavities, and then add the strapping and rigid foam as in steps 3 through 9 under "Rigid Foam Between Strapping."

ensure continuous coverage. Use a compatible caulk to caulk abutting joints of walls, floors and ceilings.

Installing a Secondary Wall

Another way to add insulation from inside the house is to build a secondary stud wall and insulate it with blanket insulation. In addition to a higher R-value, a place to run new wiring and a solid support for new wall finish is gained. The main disadvantage to doing this is that the wall encroaches and takes up floor space. Four inches are lost when 2x3 studs are set out 1 inch from the primary wall, but a 3½-inch deep cavity into which R-11 or R-13 insulation fits is gained.

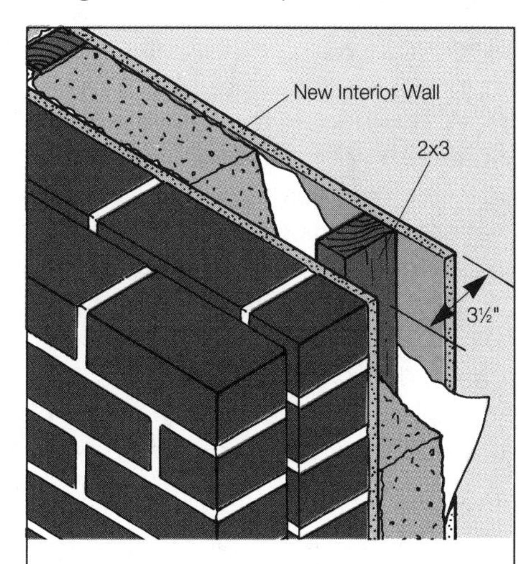

Installing a Secondary Wall. A stud wall on the inside of a wood-framed or masonry wall provides cavities for insulation and space in which to run new wiring, a vapor barrier and new wall finish.

If the primary walls are uninsulated (such as all-masonry walls) a secondary wall with R-19 added to it can be built from 2x4s and set 2 inches away from the primary wall. Before beginning, remove casings from doors and windows, electrical cover plates, and other exposed items in the primary wall. Building a new wall requires reinstalling these items in the new wall finish.

1 Attaching the Sill and Top Plates. Nail a sill (bottom) plate to the floor so that the distance from the primary wall to the outside edge of the plate is 3½ inches (for 2x3 framing) or 5½ inches (for 2x4 framing). Use 12d nails for a wood floor or construction adhesive if the floor is concrete slab. Nail a top plate to the ceiling at the same distance from the primary wall.

2 Securing the Studs. Cut two 14½-inch-long nailer pieces from the framing lumber and nail them to the top and sill plates using 12d nails. Then cut a stud to fit between the plates and toenail it into place against the nailers. Repeat the

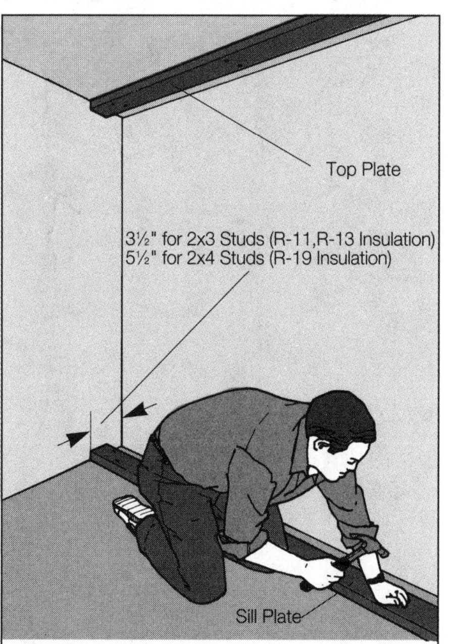

1 Attach a sill plate to the floor and a top plate to the ceiling at a distance from the wall determined by the amount of insulation to be added.

2 Cut two 14½-in.-long nailers and nail them to the top and sill plates as each stud is installed. Frame around window openings (match the line of the original rough framing) to provide support for extending the jambs and new casings.

3 Tuck 1½-in.-wide strips behind the new framing, then place full-length strips of 15-inch- wide insulation between the studs. Push each piece toward the wall, then gently pull out the edges so that it completely fills the space without creating gaps.

4 Tuck insulation around the cold side of water and waste pipes. Leave the room side uninsulated.

5 Staple 6-mil polyethylene over the face of the studs and insulation, overlapping all seams by at least 6 in. Caulk the polyethylene to abutting surfaces.

process across the length of the room until the framing is complete. Cut and fit the studs at window and door openings to match the position of the rough framing in the primary wall.

3 Installing Batts. Cut 1½-inch-wide strips of insulation and fit them into the space between the new studs and the existing wall. Then fill each bay with a length of 15-inch-wide unfaced insulation. Push each piece toward the wall, then gently pull out the edges so that it completely fills the space without gaps.

4 Fitting Around Pipes. Fit batts around water and waste pipes so that they are insulated on the cold side and uninsulated on the room side.

5 Applying a Vapor Barrier. Staple 6-mil polyethylene over the face of the studs and insulation, overlapping all seams by at least 6 inches. In places where the polyethylene meets abutting floor or wall surfaces, caulk the polyethylene to the surfaces with polyurethane or polybutylene caulk.

Controlling Moisture

Most people feel comfortable when the air they breath contains 40 to 60 percent relative humidity. Some suffer discomfort in the drier winter air. To compensate, they may keep a teakettle on top of a wood stove or run a humidifier. While additional moisture may be more comfortable, it can be a hazard to an insulated house. With escape routes sealed, the warm, moist air migrates through the ceilings and walls until it is cool enough to condense. The result is moist insulation that loses its efficiency, causing wood to decay and outside paint to peel.

There are ways to tighten up a home without suffering dry air in the winter. The natural moisture given off by bathing, showering and household plants usually is enough to ensure comfort. Here are some ways to ensure wholesome inside air while keeping moisture out of the structure:

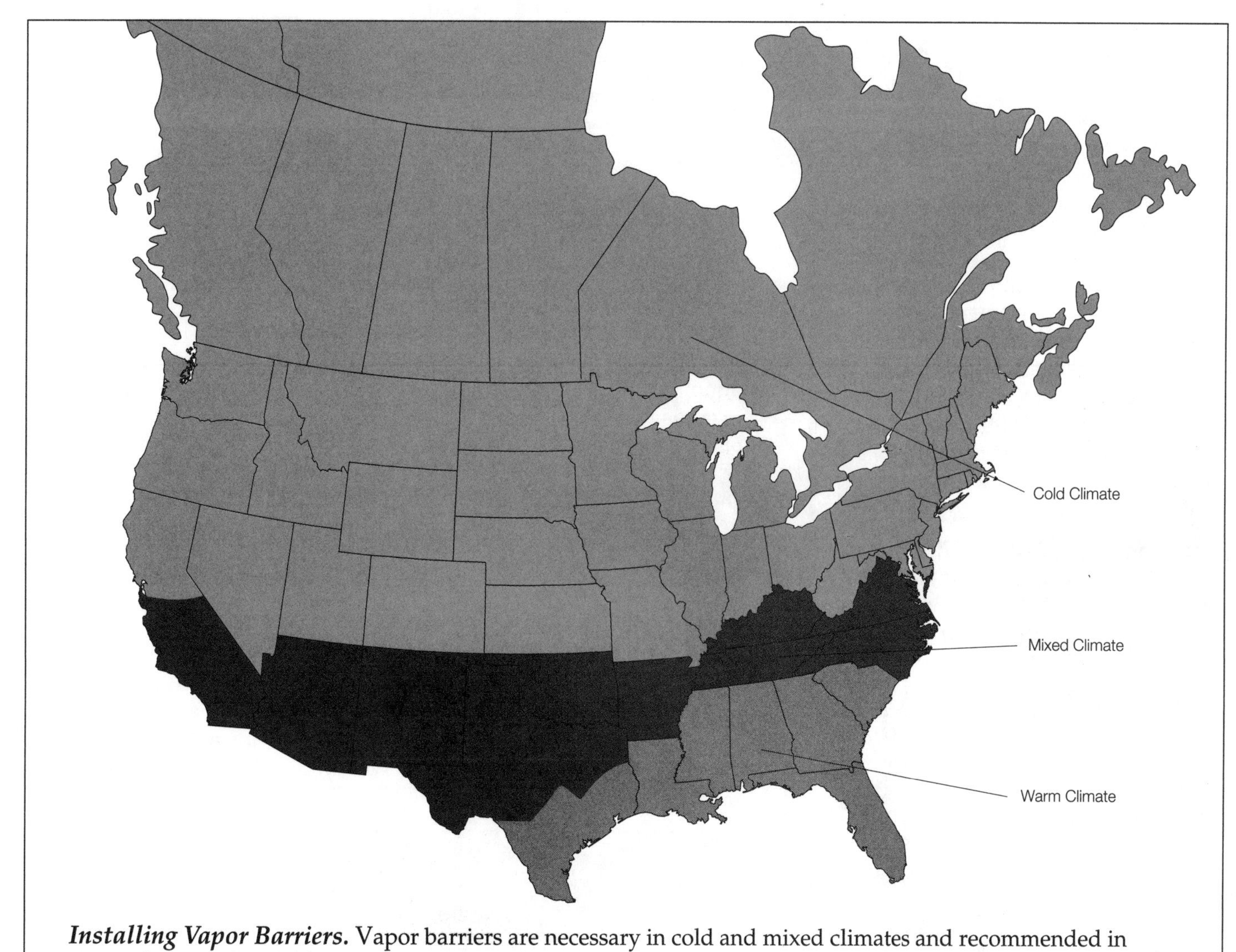

Installing Vapor Barriers. Vapor barriers are necessary in cold and mixed climates and recommended in warm climates if the house is air-conditioned.

- Do not use humidifiers to add moisture.
- Do not store green firewood inside.
- Make sure the clothes drier is vented to the outside.
- Install a vapor barrier when working with new construction or when tearing off the finish material of walls or ceilings to add insulation.
- Provide controlled ventilation in the form of individual exhaust fans or a central ventilation system.

If you find water dripping off the insides of windows during the winter, it is a good indication that the indoor moisture level is too high. An inexpensive hygrometer provides a more exact reading of the level. If the moisture level is too high, reduce it by heeding the measures suggested earlier. The next line of defense is to provide the house with adequate ventilation. Finally, use the opportunity when remodeling or adding insulation to install the vapor barriers that are effective in your particular climate.

Installing Vapor Barriers

More moisture passes through cracks and gaps by way of convection than by diffusion through the materials, so the first line of defense is sealing up airways by weatherstripping windows and doors, caulking cracks and installing air barriers beneath the siding. Once the flow of air is blocked, the passage of moisture through the materials themselves must be stopped. A vapor barrier, placed between the wall and ceiling and the side exposed to moist air, is used for this purpose. The material can be a separate sheet of moisture-impermeable material, such as polyethylene or the foil facing of insulation.

Vapor barriers are necessary in cold and mixed climates and recommended in warm climates if the house is air-conditioned for large portions of time. Two things are important when installing vapor barriers: First, the barrier must be on the proper side of the heated space (see page 38, "Cold and Mixed Climates" & "Warm Climates"), and second, it must be continuous to be effective.

Continuity is hard to achieve in existing construction. Even if the plaster or drywall is removed from outside walls, it may not be possible to extend the vapor barrier through intersecting interior walls and floors. These intersections become the weak link in an otherwise moisture-tight system. Each project is different so make the vapor barrier as continuous as possible. If it is not possible to provide a continuous vapor barrier or one that is at least nearly continuous by using plastic sheeting, the next best thing to do is to thoroughly caulk around electrical boxes, windows and doors, and the edges of new drywall.

Cold and Mixed Climates. In cold climates where moisture problems are most severe during the winter, the vapor barrier is installed on the inside of the wall.

Between cold and hot climates there is a twilight zone where both winter heating and summer cooling are about equal in importance. The best placement for a vapor barrier in this case is unclear. Some experts suggest using the same strategy used for cold climates (with the vapor barrier on the inside) because moisture problems are likely to be worse during the winter than in the summer. Most agree, however, that an air barrier (house wrap) on the outside is important to block airborne moisture.

Warm Climates. The air/vapor barrier is placed near the outside of the wall in the hot, humid areas of houses that are air-conditioned for most of the year. Check with an air-conditioning specialist or your local building department for advice on the product that best suits your home and climate. Unlike those in cold climates, crawl spaces found in warm climates must be ventilated.

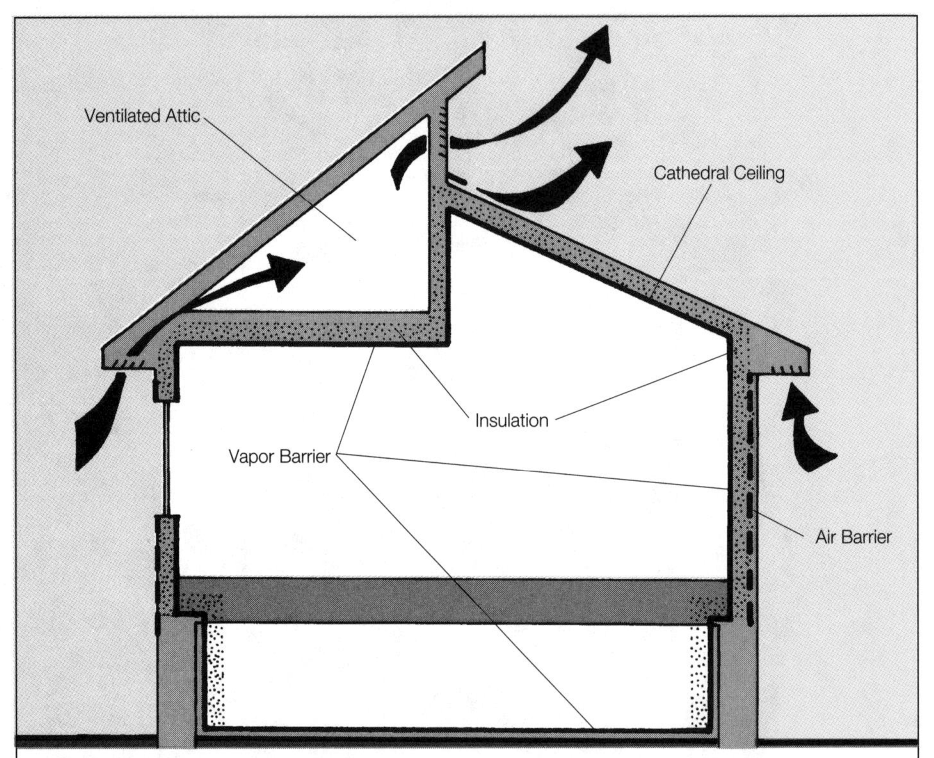

Cold and Mixed Climates. The outer shell is wrapped with an air barrier on the outside and a continuous vapor barrier on the inside. Free air circulation is encouraged through cold attics and above the insulation of cathedral ceilings, but cracks around windows and doors are sealed.

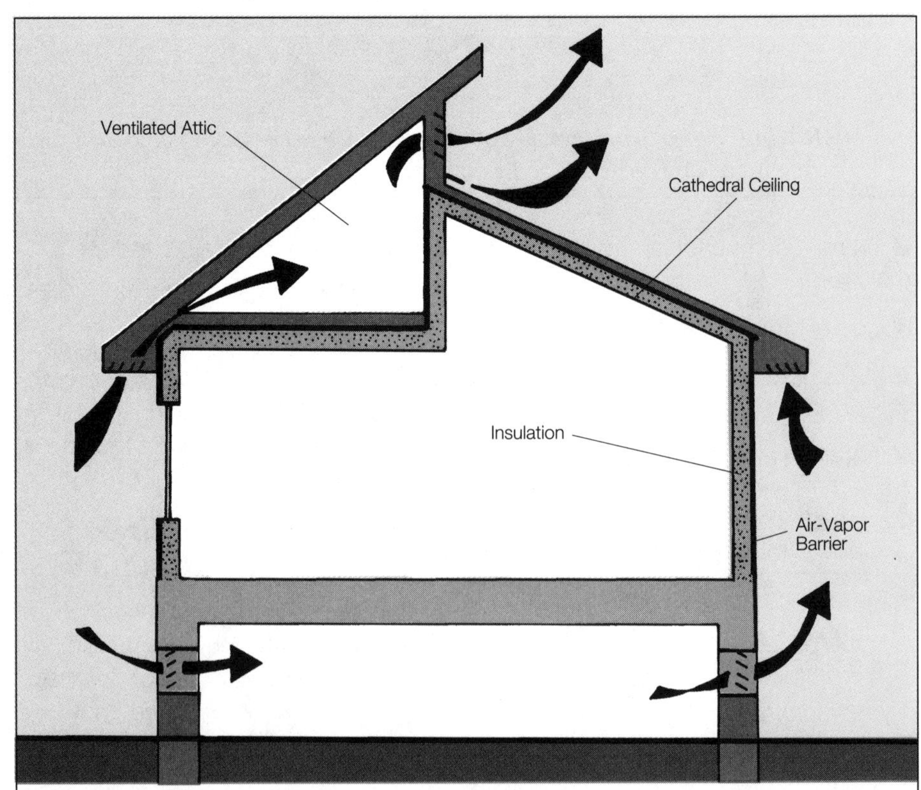

Warm Climates. Those living in hot, humid climates (and therefore using the air conditioner most of the time) benefit from placing a vapor barrier near the outside walls of their home.

Insulating Attics & Roofs

The proper way to insulate the topmost part of your home depends on whether there is an attic, and if so, whether it will be heated for use as living space. Converting attic space to a living area usually requires insulating between the rafters and side walls. The same applies to rooms that have cathedral ceilings, wherein the roof and ceiling are one. If your home has an attic, but it is used only for storage, the easiest and best approach is to insulate the attic floor.

Ventilating the Roof

Whether the attic space is insulated from above or below, adequate ventilation coming from the underside of the roof deck must be provided. Ventilation allows moisture to escape, preventing ice buildups on the eaves in winter and an overheated roof in summer. Ice buildups deteriorate the structure and cause water leaks inside the house, while dark-colored organic roofing, such as composition shingles, does not last as long if subjected to temperature extremes.

Ventilating a roof requires making a path for air to get into and out of the space between the roof deck and the insulation. The best way to insure air movement through the roof is to locate the intake port at a low point on the roof and the exhaust port at a high point.

Incoming Air. The most effective way for air to enter the roof is through vents that are placed in the soffits (which are found at the lowest points on the roof). Continuous soffit strip vents provide the most reliable port for intake air, while rectangular vents are next on the list. Round ventilator plugs are easy to install, but usually are too small to provide adequate airflow. Insulation baffles ensure a clear pathway for air to travel between the rafters.

Outgoing Air. Stale air escapes through the top of the roof through gable vents on the end walls, turbine vents on the roof, or ridge vents, also found on the roof. Continuous ridge vents are the preferred type for pitched roofs, while gravity vents are best installed where ridge vents are not feasible, such as in a hip roof. For houses that have open attics and insulated attic floors, vents located in the gable ends may suffice if the openings are large enough. Gable vents also can be used in a finished attic above an insulated flat ceiling.

Ventilating Unheated Attics

Unheated attics are the easiest type to ventilate. Air enters through vents in the eaves, rises naturally toward the roof deck, and exits through a continuous vent on the ridge, gable vents on the end walls, or roof vents placed high on the roof. If an airway is maintained under the roof (where the ceiling insulation abuts the eaves) the moving air keeps the roof ventilated, but does not affect the insulated ceiling below.

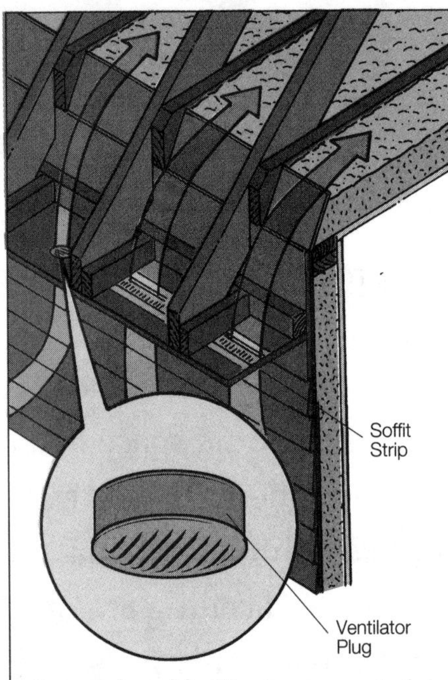

Incoming Air. The best way to let air into the roof is through soffit vents. Ventilator plugs, continuous soffit strip vents and rectangular vents are most often used.

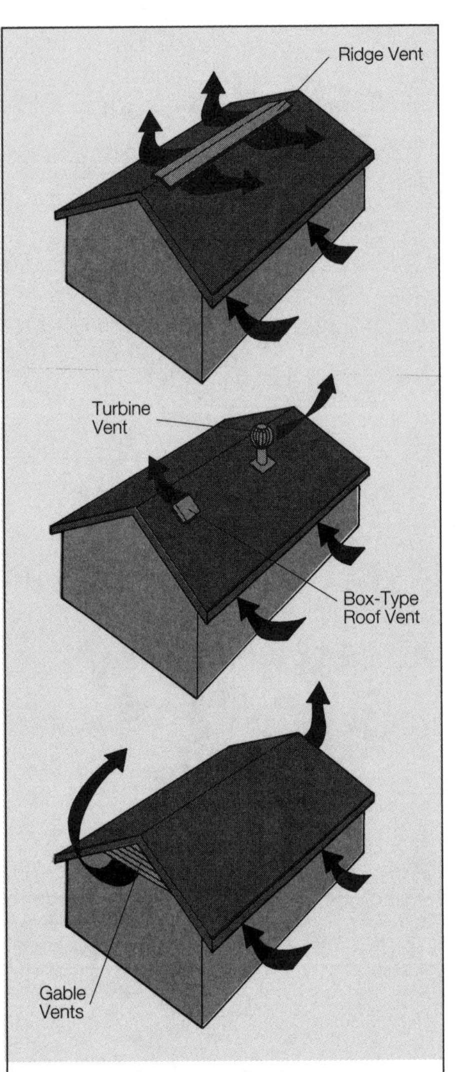

Outgoing Air. Air exhaust options include ridge vents, turbines, box-type vents and gable vents.

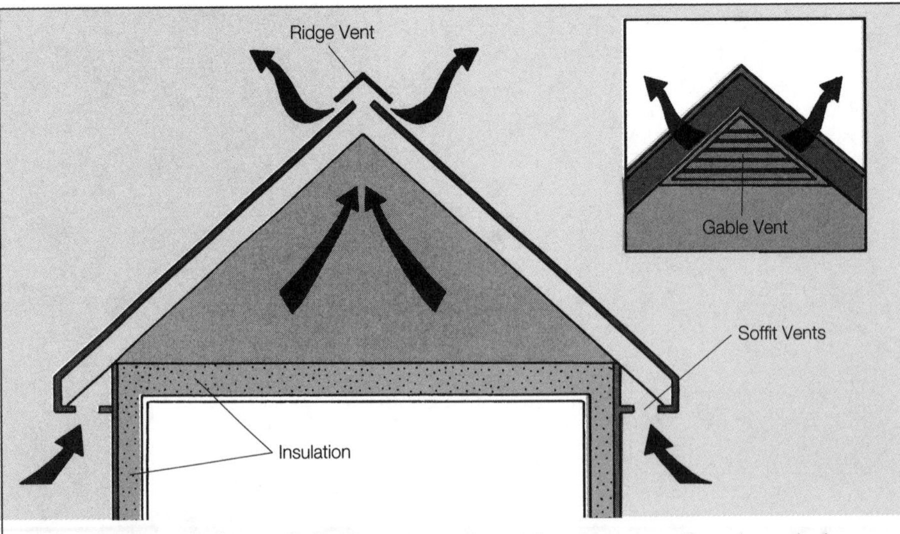

Ventilating Unheated Attics. An unheated attic is ventilated easily by installing soffit vents for air to come in and a ridge vent, gable vent or roof vent for air to escape.

Calculating Adequate Roof Ventilation

The amount of area needed to provide for roof ventilation depends on whether or not the ceiling has a vapor barrier. If there is a vapor barrier allow one square foot of free ventilating area for every 150 square feet of house area. If there is no vapor barrier double this amount. The total free ventilating area must be divided equally between the intake and exhaust ports. Because the thickness of vanes or wires in a venting device reduces the airspace, use oversized vents to compensate. For example, if the vent has a screen with 1/8-inch-square holes, divide the area of the screen by 1.25 to get the free ventilating area. If the vents used consist of louvers backed by screens, divide by 2.25. The following example shows how to size soffit and ridge vents for a 25x40-foot house that has a vapor barrier in the ceiling:

1. Figure the area of the roof: 25 multiplied by 40 equals 1000 square feet.

2. The free ventilating area equals 1000 divided by 150, or 6.66 square feet. To convert feet to inches, multiply 6.66 by 144 to get 959 square inches. You need half of this amount (480 square inches) for the soffit vents and the other half for the ridge vents.

3. Each 8x12-inch louvered soffit vent used has an area of 96 square inches. Divide this number by 2.25 to get 43 square inches (rounded off) of free area. The number of vents needed is 480 square inches divided by 43, which is 11. Use five or six on each side of the roof.

4. The product literature that comes with the ridge vent provides information concerning the necessary free ventilating area per lineal foot. One rolled ridge vent product yields 17 inches, so the house in this example needs a strip equal to the free area required at the ridge: 480 square inches divided by 17 equals 28 feet.

Ventilating Heated Attics

Providing ventilation in a heated attic is a bit trickier than working with an unheated one, but the same principles apply. The important thing is to maintain at least 2 inches of clearance between the underside of the roof deck and all insulation. This allows the air to travel from the eaves to the top of the roof. If the eaves provide a space of 10 inches or more, simply choose insulation that has a thickness that allows for this additional 2-inch airspace. Most likely though, a combination of blanket insulation between the rafters and rigid insulation attached to the underside of the rafters will be necessary to achieve the desired R-value, while retaining an adequate amount of airspace.

Ventilating Cathedral Ceilings

Like the roof deck above heated attics, cathedral ceilings must contain an airspace above insulated cavities. Unlike attics, however, not

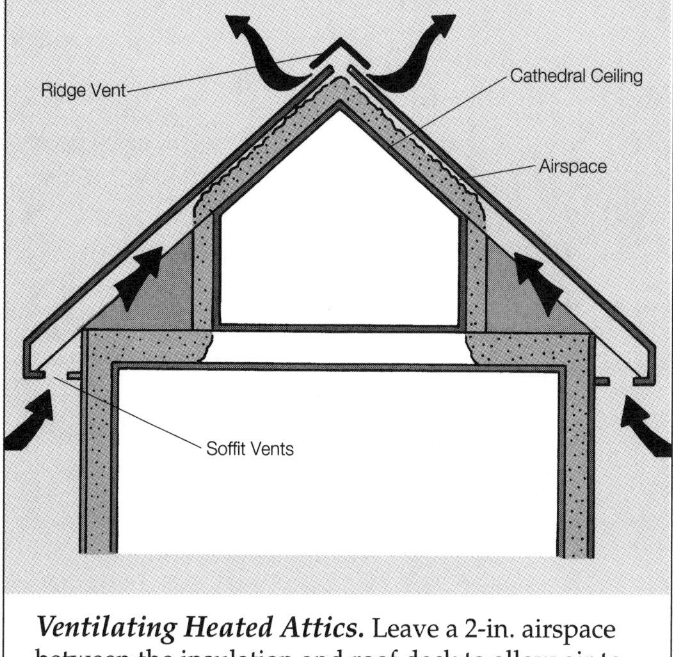

Ventilating Heated Attics. Leave a 2-in. airspace between the insulation and roof deck to allow air to circulate from the intake to exhaust vent.

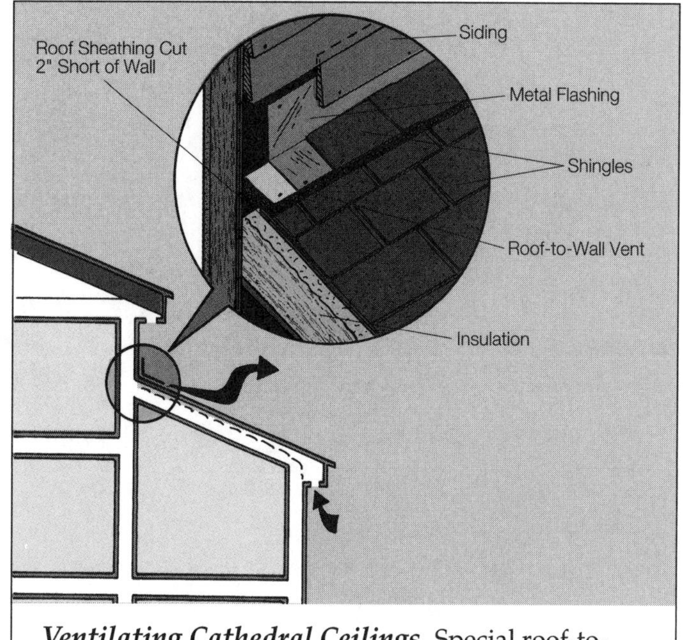

Ventilating Cathedral Ceilings. Special roof-to-wall vents provide ventilation in places where the roof abuts a wall.

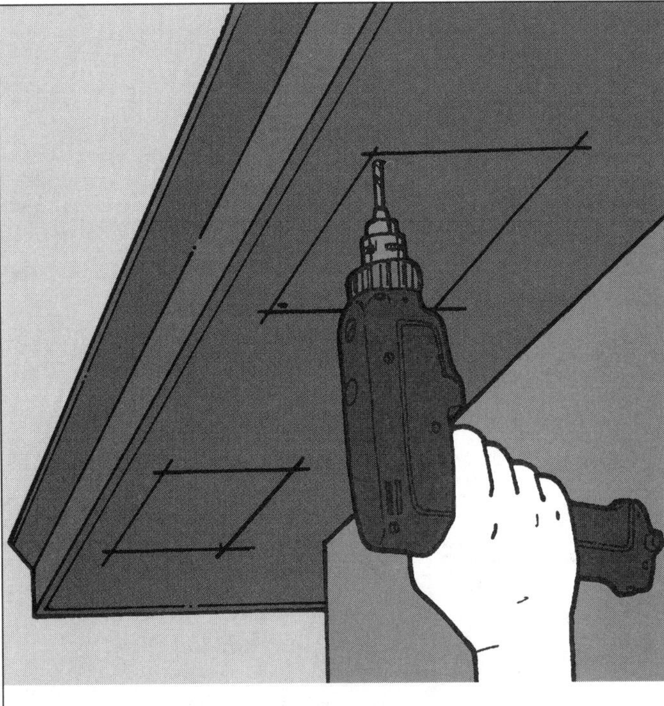

1 Use the vent as a template to mark the cutouts. Use a 3/4-in. drill bit to drill a starter hole at one corner of the marked box.

2 Insert the blade of a saber saw into the starter hole and cut the opening.

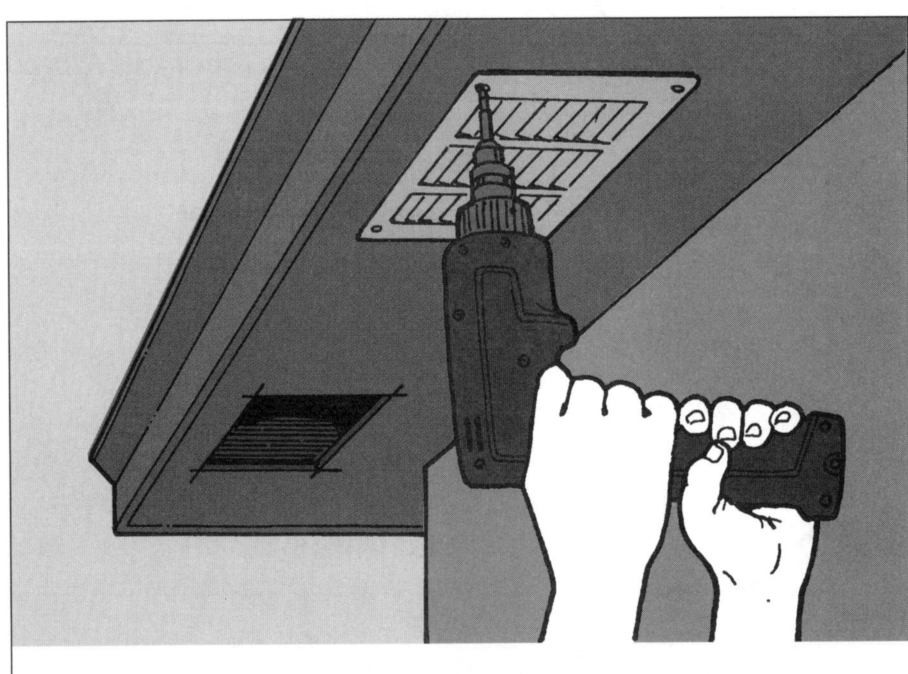

3 Use an electric drill equipped with a Phillips bit to screw the vent to the soffit.

every cathedral ceiling has an easy escape port such as a roof ridge. If not, other ways must be designed to allow air to escape. Special roof vents are available for cathedral ceilings and almost any other special conditions, such as a low roof that abuts a vertical wall.

Choosing a Soffit Vent

Continuous soffit strip ventilators work well and are easy to install into a new soffit. However, installing them into an existing soffit often requires tearing the soffit apart. Two better options in this case include rectangular roof vents and ventilator plugs.

Installing Rectangular Soffit Vents

Those who decide to use rectangular soffit vents can use the guidelines found on page 41, "Calculating Adequate Roof Ventilation," to determine the size and quantity needed at each soffit.

1 **Drilling a Starter Hole.** Mark the location of the vent on the soffit and, using a 3/4-inch drill bit, drill a starter hole at one corner of the marked rectangle.

2 **Cutting the Opening.** Insert the blade of a saber saw into the starter hole and cut the opening.

3 **Attaching the Vent.** Secure the vent to the soffit using the screws that are supplied with the vent. An electric drill equipped with a Phillips bit makes the job a lot easier.

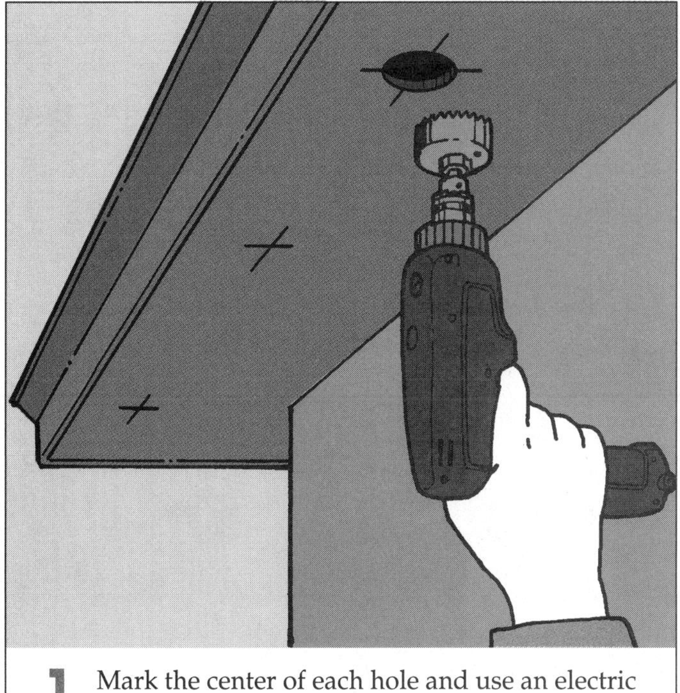

1 Mark the center of each hole and use an electric drill equipped with a hole saw bit to cut a 2½-inch diameter hole.

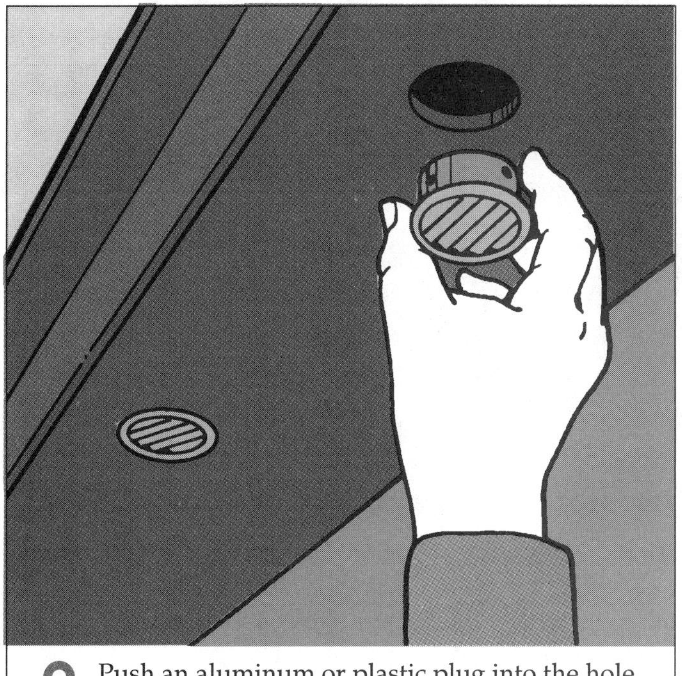

2 Push an aluminum or plastic plug into the hole until the flange is against the soffit.

Ridge Vents

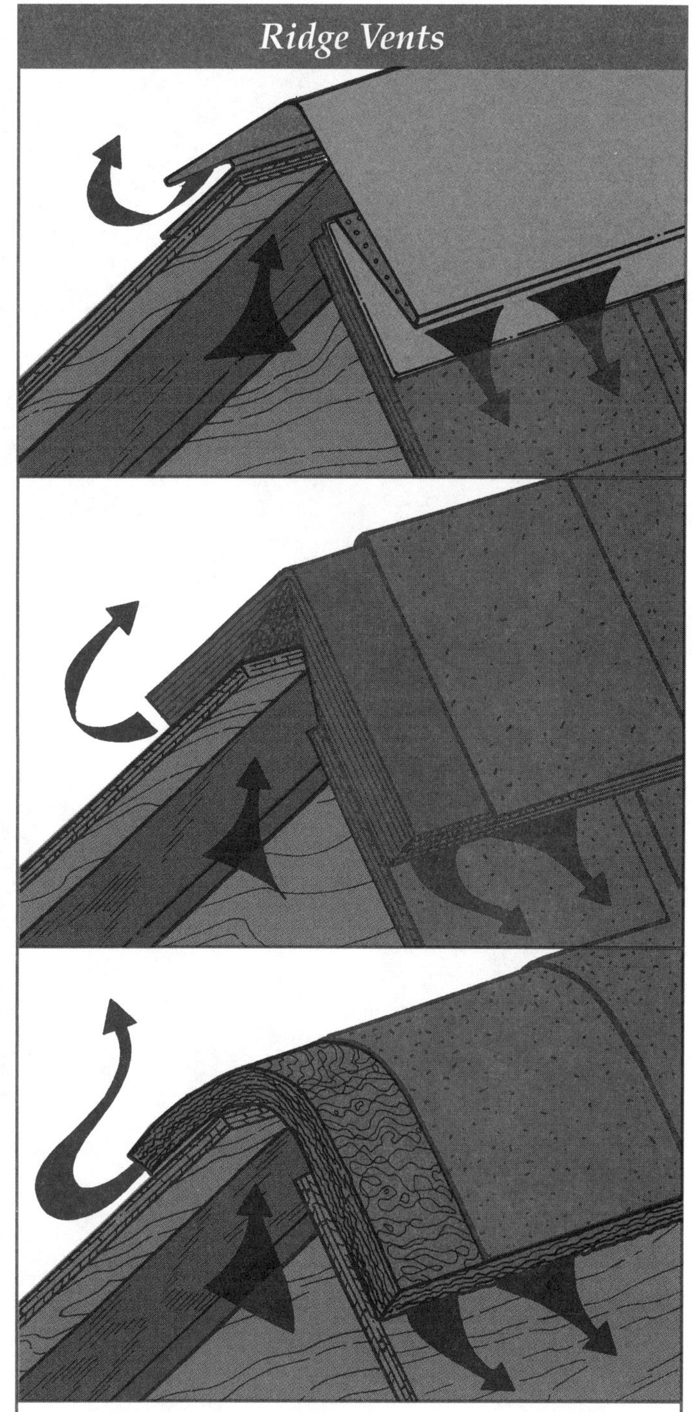

Ridge vents come as rigid lengths of metal (top), or composite materials (middle), or in rolls of an air-permeable material (bottom). The lower two are designed to accept roof shingles.

Installing Plug Vents

Ventilator plugs usually do not yield the free area required by the guidelines on page 41, but it is better to use them, than to have no ventilation at all. Use at least two plugs per rafter bay.

1 Cutting Holes. Each rafter bay requires two holes. Mark the center of each hole and use an electric drill equipped with a hole saw bit to cut a 2½-inch-diameter hole. If you do not have an electric drill, mark off the outline of the hole, drill a starter hole on the edge, and use a keyhole or saber saw to cut the hole.

2 Inserting the Plug. Push an aluminum or plastic plug into the hole until the flange is flush with the soffit.

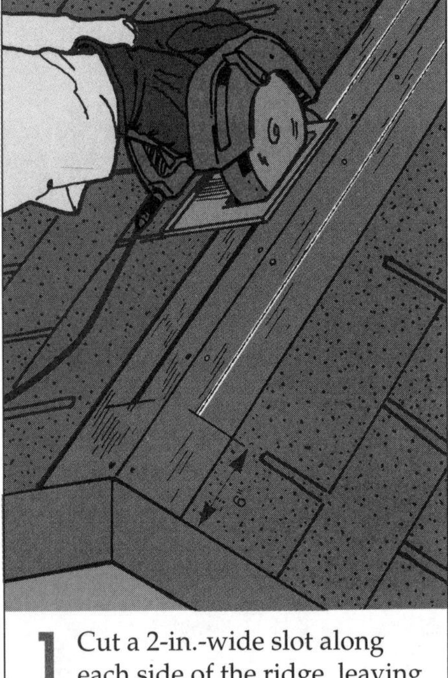

1 Cut a 2-in.-wide slot along each side of the ridge, leaving 6 in. of each end uncut.

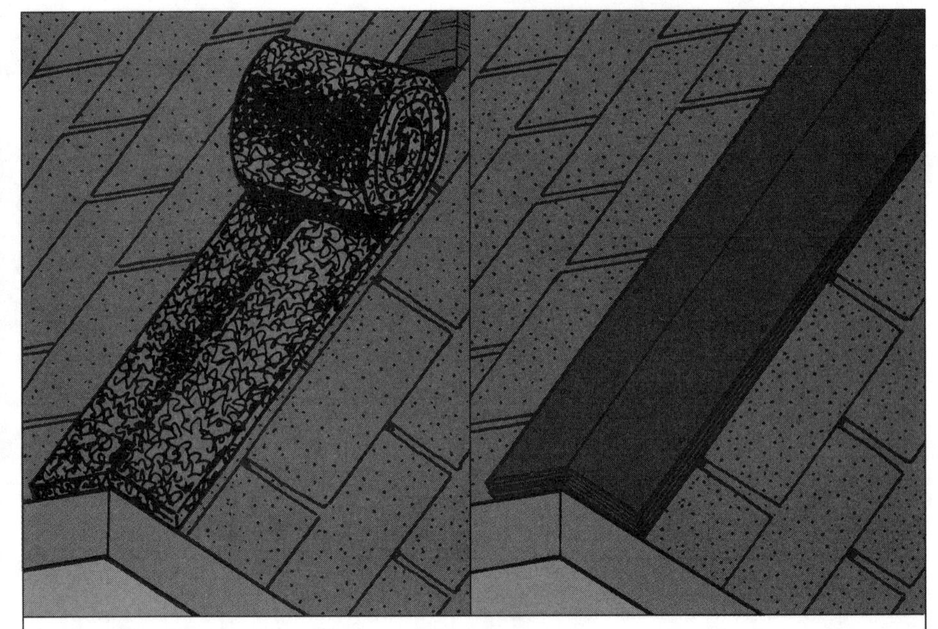

2 Uncoil a rolled vent (left) or secure a rigid vent (right) along the slot and over the uncut ends of the roof.

Installing a Ridge Vent

Ridge vents are made of various materials and may be packaged in rigid sections or rolls. A well-designed vent allows air to escape, but keeps out rain and snow. Because the ridge vent sits on top of the roof, it is highly visible. Keep this in mind when choosing a vent. Many homeowners opt for one that can have shingles installed on top.

1 Cutting an Opening. Use a circular saw with a carbide-tipped blade to cut a 2-inch slot along each side of the ridge. Start the cut 6 inches from one end of the ridge, and end it 6 inches from the other end. Set the blade depth to cut through the roofing and sheathing only, leaving the rafters uncut.

Caution: *Wear goggles and make sure you have solid footing when cutting with a saw. Install roofing cleats if necessary to support yourself on the roof. Lay out the extension cord so that there is no chance of tripping on it.*

2 Attaching the Vent. Uncoil the vent (if packaged in rolls) or secure a rigid vent to the ridge. Extend the vent to cover the uncut 6-inch portions of roof at each end.

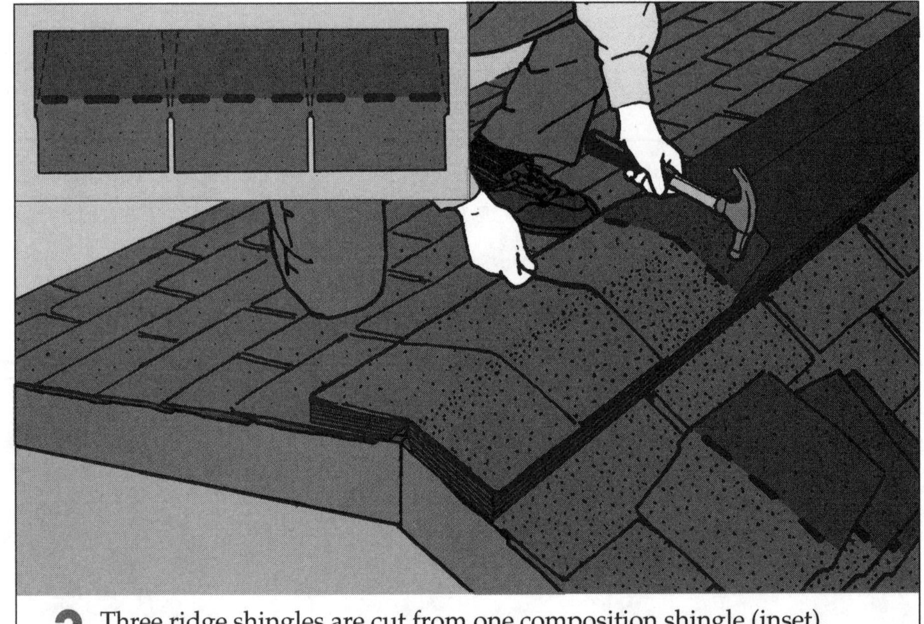

3 Three ridge shingles are cut from one composition shingle (inset). Place ridge shingles over the top of the vent. Without compressing the vent material, drive the nails through the shingles and vent and into the sheathing.

If using a metal vent that is not intended to be topped with shingles, nail the flanges to the roof with aluminum roofing nails.

3 Topping the Vent. For ridge vents that are designed to be covered with shingles, cut three ridge shingles from one three-tab composition shingle. Place the first shingle over one end of the ridge vent, aligning it with the edge of the shingles along the roof rake. Nail the shingle through the part to be covered by the next shingle. Using roofing nails that are long enough to penetrate into the sheathing, install one nail at each side of the ridge. Put the next shingle in place over the nails and continue.

Installing a Gravity Roof Vent

While ridge vents are the preferred way to create an escape outlet near the top of a roof, they are not easy to install on some houses. For example, ridge vents are not suited to houses that have hip roofs. Instead, one or more gravity vents usually are installed.

Turbine vents are a type of gravity vent that have vanes located in the turbine-shaped top of the vent, which usher air to the outside. When there is a breeze the vanes move, increasing the capacity of the device by an average of 130 percent. The downside to turbine vents, other than their prominent appearance, is a tendency to ice up in winter. Some homeowners prevent this by cloaking them in garbage bags in winter; a measure that prevents them from working when they are most needed.

Box-type gravity vents do not have moving parts. They have louvers on all sides but the top. The step-by-step instructions for installing them are the same for turbine vents.

1 Cutting an Opening. Drive a nail through the roof sheathing from the attic below to locate a clear opening between framing. Using the nail as a guide, mark an opening the same size as the throat opening of the vent. Then remove the nail and use a carbide-tipped saw blade to cut out the opening.

Caution: *Wear goggles and make sure you have solid footing when cutting with a saw. Install roofing cleats if necessary to support yourself on the roof. Lay out the extension cord so that there is no chance of tripping on it.*

2 Adjusting the Roofing. Use a utility knife that has a hooked blade to cut and remove shingles so that the top of the roof vent slides under the two topmost shingle courses. Use a pry bar to pull out nails that may interfere when sliding

1 Drive a nail through the roof sheathing to determine where to make the cut. Mark an opening the same size as the throat opening of the vent. Use a carbide-tipped saw blade to cut the opening.

2 Cut and remove enough shingles so that the top of the roof vent can slide under the two topmost shingle courses. Use a pry bar to pull out nails that block the path of the vent. Apply roofing cement around the cut opening and under the point at which it will contact the flange.

3 Slide the vent under the top two shingle courses and over the bottom two. Lift up the shingles and use galvanized or aluminum nails to nail through the flange around the vent.

4 Apply roofing cement under each shingle and press it firmly into place. Apply a bead of caulk around the joint between the shingles and the vent.

the vent under the courses. Before placing the vent, apply roofing cement around the cut opening and under the point at which it will contact the flange.

3 Installing the Vent. Slide the vent under the top two shingle courses and over the bottom two courses. Then lift up the shingles and, using nails made of the same metal as the flange, nail the vent through the flange.

4 Sealing the Roof. Apply roofing cement using a stick from a gallon-sized container, or better, a cartridge in a caulking gun. Press each shingle firmly into place. Finally, lay a bead of caulk around the joint between the shingles and the vent.

Insulating an Attic Floor

Note: *When working in the attic (or a confined space where light is limited) it is a good idea to hook up a portable light. Its installation is as simple as tacking a nail into a rafter. Just make sure the cord does not lay where you can trip over it.*

If the attic does not have to be heated, insulate the floor rather than the roof. Insulation can be placed directly over an existing subfloor if it is not needed for storage. If there is no floor surface, insulate between the joists. If preferred, an insulation contractor can be hired to blow loose-fill insulation into the space.

Decide on an R-value and choose insulation of a thickness that allows the installment of two layers. The top layer runs across the bottom layer, thereby sealing the joints. For example, if 4 inches of mineral wool insulation already exists between the joists, the attic floor has an R-value of about 15. To bring the R-value up to R-38, an extra R-23 is needed. One layer of R-11 plus one layer of R-13 blanket insulation meets the goal. Buy enough rolls of unfaced insulation in these thicknesses to cover the attic floor.

Sealing an Attic Hatch

Though the access hatch to the attic is seldom used in an unheated attic, it must be well insulated to avoid subverting the hard work and gains made by insulating the attic floor. If an access panel or door already exists, glue down two or three layers of rigid foam to the top of it. Then, to prevent drafts, weatherstrip the stops below.

If the hatch does not create an airtight seal, improve it or build a new one from scratch. All necessary work is easier done before, rather than after, the attic insulation is installed.

Building a 2x4-foot hatch requires basic carpentry tools and the following materials (if the opening is larger than 2x4 feet, increase the quantity of materials accordingly): one sheet of 1/4-inch plywood; two 2x4-foot sheets of 1½-inch-thick rigid foam; two 12-foot-long 2x4s; a pair of 3x3-inch hinges; one door pull, and about 9 feet of compression-type weatherstripping.

1. Preparing the Opening. Mark a line 4½ inches down from the attic floor, around the inside of the hatch opening. Cut jamb pieces from the 2x4s and nail them around the edge so that the tops of the jambs meet the line.

2. Making the Hatch. Measure the width and length of the opening and cut two pieces of plywood to measure about 3/4 inch less in both directions. Cut 2x4s to lengths that fit around the inside of the plywood, and nail them to one sheet using 6d nails spaced 8 inches apart. Use a kitchen knife to cut two pieces of rigid foam insulation and fit them snugly inside the core. Nail the top sheet of plywood to the 2x4 edging, then screw the hinges to one side of the top. Mount door pulls on the front and back at the opposite edge from the hinges.

3. Mounting the Hatch. From the attic side, screw the hinges to the attic floor. Then go below and close the hatch. Working from the underside with the hatch closed, tack weatherstripping around the jamb pieces to fit snugly against the underside of the hatch.

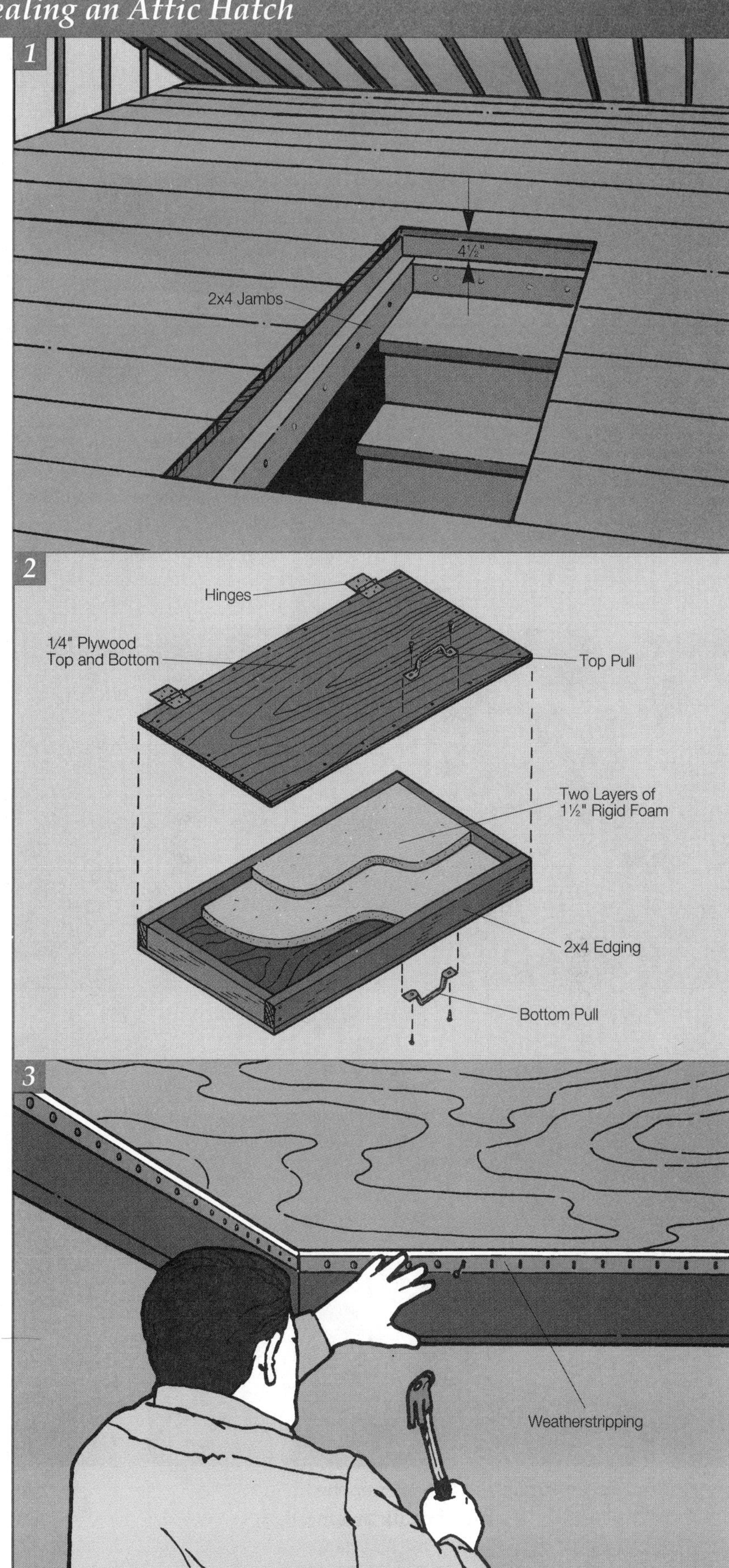

Insulating over the Flooring

Placing insulation directly over the top of the attic floor is quick and easy, but has two drawbacks: The floor cannot be used for storage, and there is no chance to install a vapor barrier. If the room below does not have a vapor barrier, consider ripping up the attic floor and installing one as described on page 51, "Installing a Vapor Barrier above the Floor." An alternative to this is to install a vapor barrier under a new ceiling finish. The following method shows how to insulate above an attic floor using unfaced blanket insulation.

1 Prying off a Strip of Flooring at the Sides. If the floor has not been insulated, or if the insulation does not meet the underside of the floorboards, fill in the gap around the perimeter before adding insulation above the floorboards. If this is not done, the airspace trapped under the floor acts like a short circuit by providing a direct path for air to circulate and escape through the outside edges. Pry up floorboards that are within 18 inches of where the rafters meet the floor.

2 Cutting a Strip out of the Ends. At opposite ends of the attic floor where the boards run lengthwise, use a circular saw to cut away one bay of flooring. This cut is made flush with the first floor joist in from the end.

3 Filling in the Perimeter Strip. After removing the flooring from the outer perimeter of the attic floor, it is easy to fully assess the depth of the existing insulation and estimate its R-value. If some insulation is present, but it does not fill the full depth of the floor joists, top it off with loose-fill insulation, such as cellulose or mineral wool. If there is no insulation at all, fill the void with loose-fill, or cut pieces of blanket or batt insulation and fit them between the joists. Make sure that cavities are filled all the way to the top of the joist.

4 Installing the First Layer. Using blanket or batt insulation, start at one edge of the attic floor and roll out the first layer so that it runs lengthwise and parallel to the floorboards. Press the insulation so that each piece is snug with the abutting rafters. Leave a space between the soffit and the roof deck to allow air to circulate through the soffit vents. (If there are no soffit vents, install one of the vents described on page 42, "Choosing a Soffit Vent.")

1 Pry up floorboards that are within 18 inches of where the rafters meet the floor.

2 Use a circular saw with a carbide-tipped blade that is set to the depth of the flooring to cut crosswise through the floorboards.

3 Pour loose-fill insulation from a bag around the perimeter of the attic floor until it is filled. Level each pour even with the top of the joists.

4 Start rolling blanket or batt insulation from the eaves side, working toward the center of the attic floor. Leave a 2-in. airspace at the eaves for air to circulate from the soffit vents into the attic.

5 Cut the insulation to fit snugly against the gable wall. If necessary, cut the insulation to fit between studs.

6 Roll out the second layer perpendicular to the first. When you reach the attic hatch go to the opposite wall and complete the installation. Leave the work adjacent to the hatch for last.

5 Fitting the Ends at the Gable Wall. Use a serrated knife to cut the ends of the blanket. Tuck the cut ends into the bottom of the stud bays at the gable wall.

6 Putting down the Second Layer. Beginning at the gable end, cut pieces for the second layer to fit between the studs. Tuck them in place, then roll out blankets to run perpendicular to the first layer of insulation. Fit the ends of the pieces between rafters, but again, be sure to leave space for the soffit vents. You have to walk on top of the first layer, so make sure the strips fit together snugly.

Caution: *Remember that there are no floorboards around the perimeter of the attic, so do not step there.*

Insulating Between Attic Joists

If the attic has no floor and no insulation, the spaces between joists can be insulated. Either hire an insulation contractor to install loose-fill insulation, or do the job yourself by installing blanket or batt insulation. If there is no vapor barrier between the attic joists and the ceiling below, install one between the new insulation and

Working Around Recessed Lights

Incandescent lamps give off a lot of heat. If this heat is trapped below ceiling insulation it poses a fire hazard. For this reason, the National Electrical Code requires a 3-inch gap around the fixture (left), or the use of a type of fixture that is rated for an insulated cover. Both solutions have drawbacks. Building a box around the fixture allows heat from the room—as well as the lamp—to escape through the ceiling. Lamps rated for an insulated cover have the annoying habit of shutting off automatically when they get too hot.

One way out of this dilemma is to replace the fixture entirely with one mounted to the surface of, rather than recessed into, the ceiling (right).

the heated space below (either put down polyethylene from the attic side as described on page 51, "Installing a Vapor Barrier above the Floor," or put polyethylene over the ceiling below and cover it with a new ceiling finish material).

This job requires working above exposed joists without the benefit of a solid surface upon which to walk. Use wide boards or pieces of plywood placed over the tops of joists to create a working platform. Move the pieces along with you as you work. Be careful not to step or lean on the edges of the boards. They could tip up, resulting in an ankle sprain or a foot through the ceiling.

Plugging Gaps in the Ceiling

Before filling the joist spaces with insulation, all air leaks around pipes, wires, chimneys and electrical boxes must be plugged.

Sealing Around Chimneys. Building codes require a 2-inch airspace between combustible wood framing and the chimney. To prevent the gap from losing heat from below, cut lengths of aluminum flashing or sheet metal to size. Then, with one edge in contact with the chimney, fit them around the top of the framing. Use aluminum roofing nails to secure the metal to the framing. Finally, use a caulk that is designed for high temperatures (such as neoprene or polysulfide) to caulk the joint between the chimney and metal.

Sealing Electrical Boxes. Use pieces of polyethylene sheet to make shrouds to fit over electrical boxes that are mounted on the ceiling. Use a caulk (such as polyurethane or polybutylene) to attach the shrouds to the framing and ceiling. Do not put a shroud over recessed light fixtures. Seal or replace them as described on page 49, "Working Around Recessed Lights."

Sealing Around Pipes and Wires. Wrap pipes and wires with pieces of polyethylene. Secure the polyethylene to the ceiling (or vapor barrier sheet) using polyurethane or polybutylene caulk. Tape the top tightly to the object.

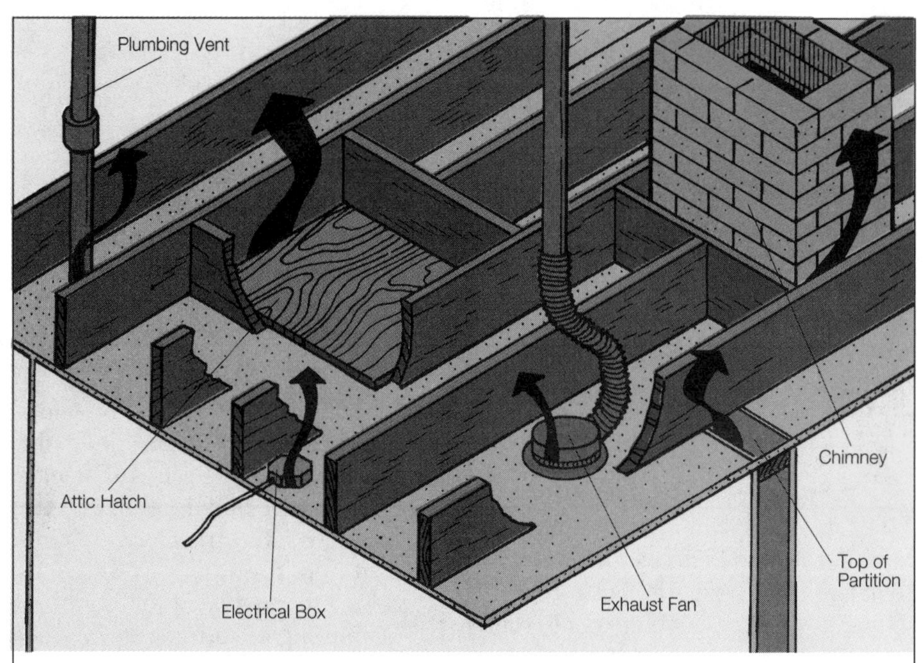

Plugging Gaps in the Ceiling. Before filling the attic floor joists with insulation, check out all sources of air leaks and take the appropriate means to seal them.

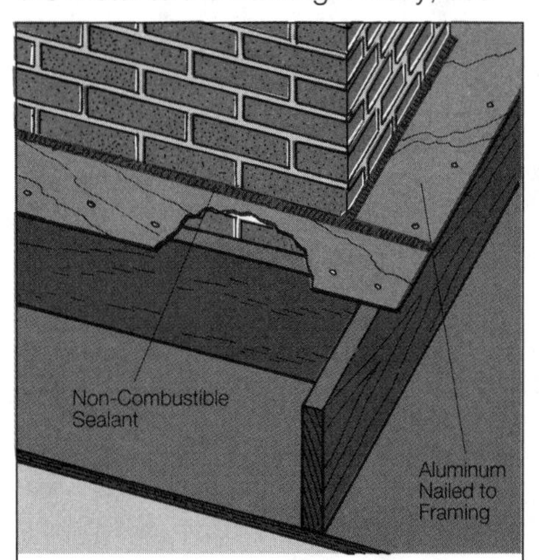

Sealing Around Chimneys. Cut and fit aluminum flashing around the framing. Use a noncombustible caulk to close the gap between the metal and the brick.

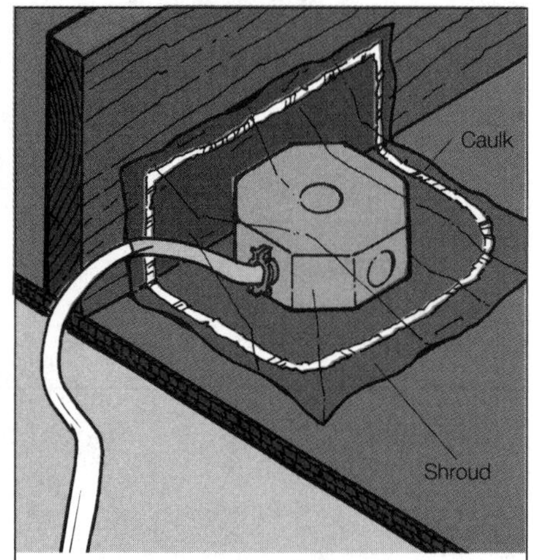

Sealing Electrical Boxes. Cover the protruding backsides of electrical boxes with shrouds made of 6-mil polyethylene caulked to the substrate.

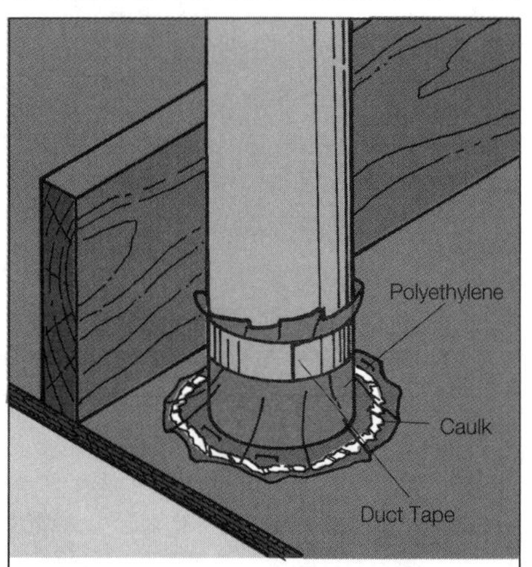

Sealing Around Pipes and Wires. Wrap protruding pipes and wires with polyethylene and secure with duct tape and caulk.

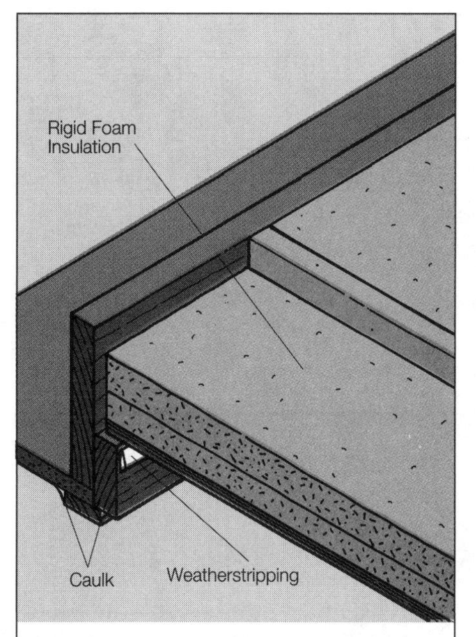

Tightening an Access Panel. Insulate an existing access hatch by attaching rigid foam to the attic side of the panel. Add caulk and weatherstripping.

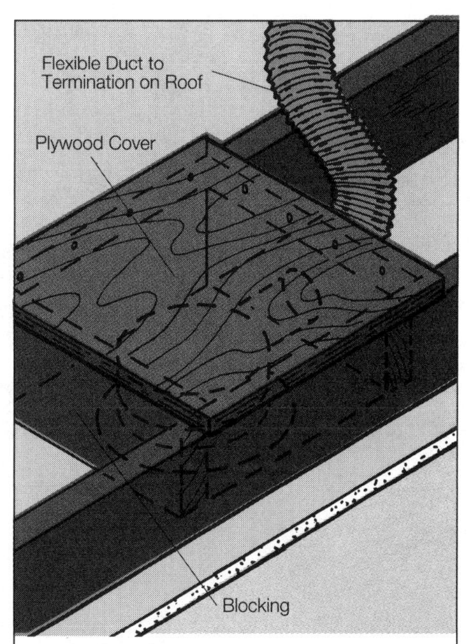

Enclosing Exhaust Fans. Seal the top side of an exhaust fan by installing blocking between the joists and attaching a piece of plywood to the top.

Tightening an Access Panel. A loose-fitting access panel loses heat through the panel itself, especially around the edges. To build an appropriate access hatch see page 47, "Sealing an Attic Hatch." To tighten up an existing hatch, attach rigid foam insulation to the attic side, install weatherstripping around the jamb, and caulk wood trim joints. Use your finger to smooth the caulk, and then repaint. Figure on at least 3 inches of thickness for the foam. To attach the rigid foam insulation, use an adhesive that is specially made for foam plastics and is available in caulking cartridges. Take care to prevent the adhesive from going into the joint where it may keep the hatch from opening.

Enclosing Exhaust Fans. First, make sure that all exhaust fans expel exhaust to the outside through a duct, and not into the attic itself. A fan that exhausts into the attic invites damage to the structure. If this is the case, the exhaust must be redirected. To do this, buy a roof or wall termination kit and flexible duct. Install the kit in the roof or a nearby gable wall, as described on page 68, "Installing a Bathroom Exhaust Fan." Then put blocking between the joists and a piece of plywood above to enclose the fan housing.

Installing a Vapor Barrier above the Floor

The most effective way to install a continuous vapor barrier is to apply it to the ceiling from the room below. There are two ways to do this. The first is to gut the old ceiling material to expose the joists, then staple the polyethylene sheet to the exposed bottoms of the joists. This is a good technique if the ceiling plaster is in bad shape and needs to be replaced with drywall anyway. The second possibility is to leave the defective plaster in place, staple a polyethylene vapor barrier directly over it, and then attach a layer of drywall to serve as the new ceiling.

However, if you do not want to alter the ceiling finish you can install a polyethylene vapor barrier to the top side of the attic floor as described here.

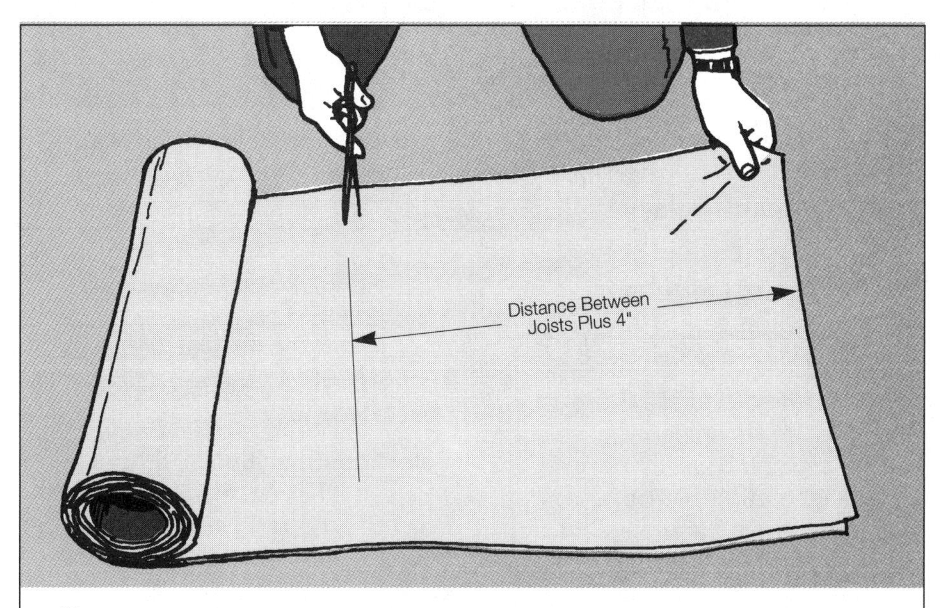

1 To simplify cutting, unroll just enough polyethylene for each strip and cut through all four layers at once. When unfolded you will have an 8-foot-long strip as wide as the joist space plus 4 in.

1 Cutting Strips. Six-mil polyethylene sheet is available in 2-foot-wide rolls of material folded into four layers. When unfolded, the sheet is 8 feet wide. To simplify cutting, unroll just enough material for each strip and cut through all four layers at once. Strips must be as wide as the distance between ceiling joists, plus 4 inches (to cover 2 inches of the side of each joist).

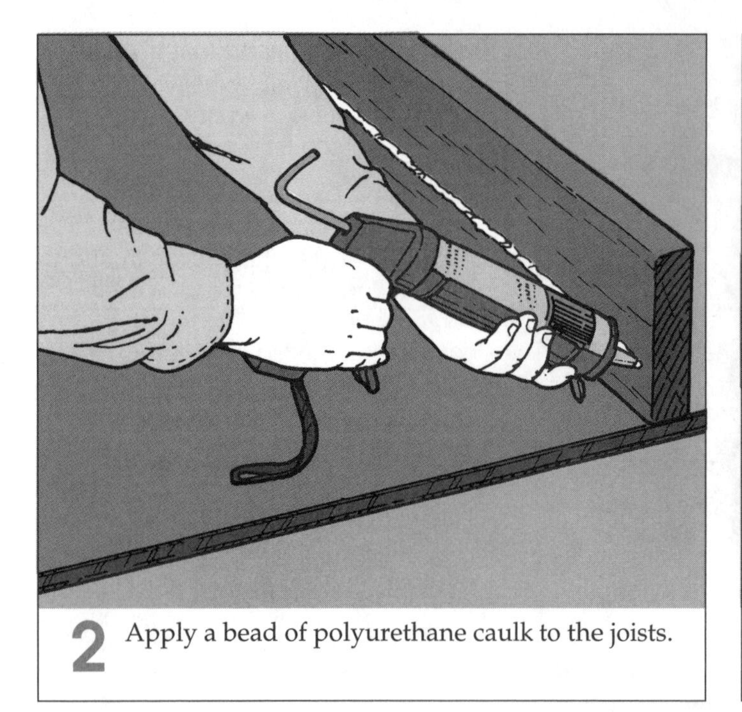

2 Apply a bead of polyurethane caulk to the joists.

3 Staple the polyethylene through the caulking bead into the joists at about 6-in. intervals.

2 Caulking the Joists. Use a caulking gun to run a bead of polyurethane or polybutylene caulk along the face of the ceiling joist at each side of a joist bay. Locate the bead about 1½ inches above the bottom of the joists.

3 Attaching the Polyethylene. After each bay has been caulked, lay in a strip of polyethylene and press the edges into the caulk. Then, using 1/4-inch-long staples, staple through the polyethylene about every 6 inches to keep the polyethylene from peeling away from the caulk.

4 Sealing the Vapor Barrier to Objects. The vapor barrier must be continuous in order for it to be effective. Cut the main sheet to fit over projections such as wiring and piping. Use a piece of polyethylene to make a shroud large enough to enclose an object, such as an electrical box, or to fit around a penetrating object, such as a pipe or wire (see page 50, "Plugging Gaps in the Ceiling"). Use polyurethane caulk and/or duct tape to seal the shroud to the main polyethylene sheet.

Caution: *Do not enclose recessed lighting fixtures; doing so creates a fire hazard.*

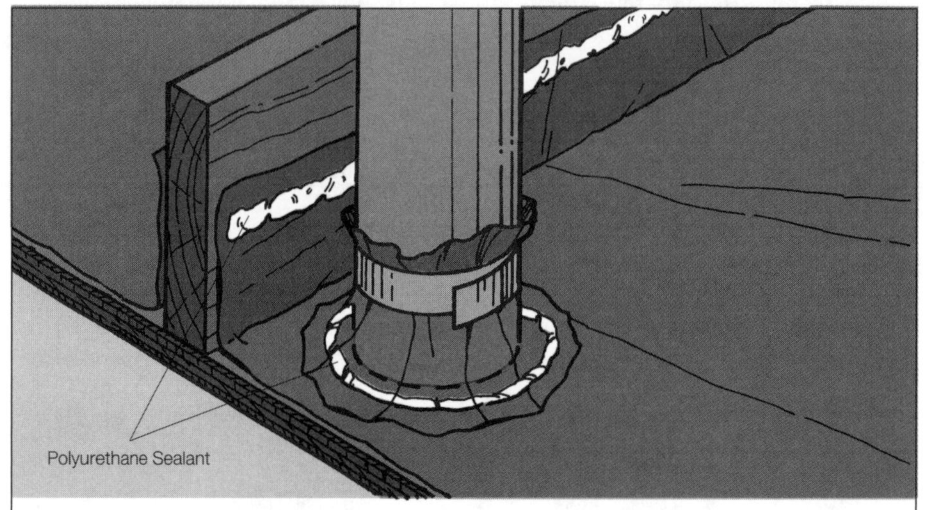

4 Cut and fit the polyethylene vapor barrier around protruding objects, and then caulk the barrier sheet to the objects or to their shrouds (if they were installed first).

Laying in the Insulation

If the attic does not have a floor and a floor is not going to be installed, there is no reason to allow joist depth to limit the amount of insulation installed. Instead, select a target R-value (see page 10 "Targeting Insulation Levels," and page 14, "Choosing the Right Insulation") and achieve it by filling the joists to the top using blanket or batt insulation. Then lay blankets or batts across the tops of the insulated joists. For example, R-25 attic insulation fills 8-inch-deep joists, but if the target R-value is R-38, the first layer can be over-laid with 3⅕-inch R-13 blankets to achieve the goal.

Selecting Insulation Width. The width and thickness of the insulation used depends on the spacing of the ceiling joists. To determine the width needed, measure the distance between two facing joists from the face of one to the opposite face of the next. If the house was built within the last 50 years or so the joists probably are set at 16- or 24-inch centers. Select the appropriate width insulation (15- or 23-inch wide) to fit snugly between the joists. Joists in older houses sometimes are spaced irregularly, such as at 32-inch intervals. In this case, buy 23-inch-wide insulation and cut it into several cross pieces to fit across rather than lengthwise.

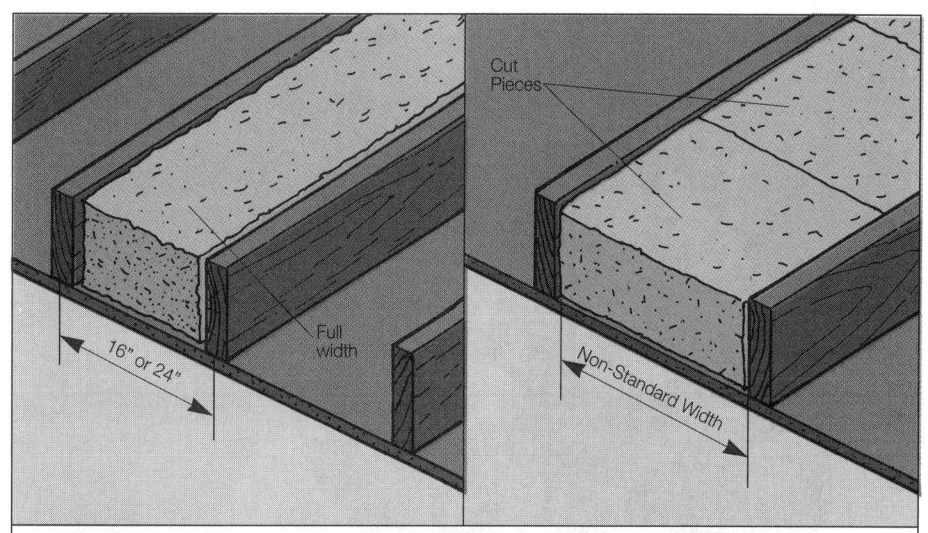

Selecting Insulation Width. If the joists are spaced at 16 or 24 ins. on center, select 15- or 23-in.-wide insulation. For wider spacings use 23-in. insulation cut in pieces lengthwise to fit.

1 Cutting the Insulation. Cut pieces of insulation that are about half as long as the width of the floor between opposing rafters. Place the blanket or batt to be cut on a piece of wood (the plywood that makes up your working platform is good for this). Place a board (a 1x6 will do) over the top of the insulation and press it lightly. Use a serrated kitchen knife to cut through the insulation.

1 Use a serrated kitchen knife held against a board to cut through blankets or batts. A piece of plywood provides a good cutting surface and working platform.

2 Fitting the First Layer. Start placing insulation at the eaves, leaving a space at the soffit to allow free air to flow from the soffit vents to the underside of the roof. Go to the opposite eaves and install a piece to meet roughly at the center. Cut the end of the second piece to allow the two pieces to meet snugly.

3 Laying the Top Layer. Starting at a gable end, unroll strips and place them across the joists, butting them against each other as you go. As with the first layer, make sure to leave a gap for airflow below the roof surface at the eaves. Plan the work as if you are painting a floor, working toward the point of exit which, in this case, is an access hatch.

Insulating an Attic Roof or Cathedral Ceiling

If the attic is used for living space, or if there is no attic above (as in a cathedral ceiling), the only choice is to insulate within the roof. The cavities between rafters or ceiling joists may or may not be deep enough to contain the desired thickness of

2 Beginning at the eaves, fit each piece into the rafter cavities. Cut pieces to meet near the middle of the floor, then complete the bay starting from the opposite eaves.

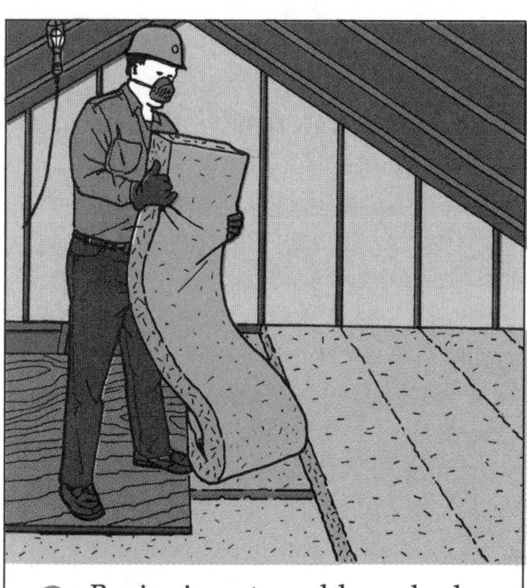

3 Beginning at a gable end, place the second layer across the first and work toward the hatchway.

1 Push each 4-ft.-long insulation baffle into place against the roof deck and use 1/2-in. staples at the edges.

insulation, while still allowing 2 inches above for air to circulate under the roof deck. If more insulation is needed, rigid foam can be added below the cavity insulation.

Installing Batts Between Rafters

The following steps show how to insulate between the rafters above an attic that will be finished, or a cathedral ceiling. Be sure that all rough wiring and electrical junction boxes are in place before beginning the insulation work.

1 **Installing Insulation Baffles.** Fitting blanket insulation into the spaces between rafters in a way that leaves a clear, 2-inch airspace under the roof deck is a hit-or-miss proposition unless insulation baffles are first installed against the roof sheathing. Made of 1/2-inch-thick polystyrene beadboard, these devices are formed into a corrugated shape that guarantees air channels next to the roof. Insulation baffles are available at most home centers and lumberyards in lengths of 4 feet and widths to match 16- and 24-inch rafter spacing. Simply place each baffle in the cavity and use 1/2-inch staples to attach them to the roof deck.

2 **Securing Batts.** Batts faced with kraft paper or foil are available in widths of 15 or 23 inches. Tabs that extend past the width of the batts are stapled along the rafters. Use 1/4-inch staples spaced every 6 inches to secure the batts. If the rafters are closer than standard width, cut batts along their length and create tabs at the cut ends by rolling back the insulation an inch or so.

If using unfaced batts, secure them with wire supports called tiger's teeth. The supports are slightly longer than the cavity width so they bow upward, pressing against the batts and digging the ends of the tiger's teeth into the sides of the rafters.

Tiger's teeth are available only for rafters spaced 16 or 24 inches on center. To support unfaced batts between other spacings, staple string in a zigzag pattern across the bottom of the rafters. If working alone, staple sections of about 2 to 4 feet at time, then insert a batt at one end and pull it through the section. The process is faster when a helper holds each batt

2 Faced insulation has paper flanges that are stapled to the rafters (top). To secure unfaced insulation use tiger's teeth (center) or zig-zagged string (bottom).

3 Staple a continuous layer of 6-mil polyethylene over the face of walls and roofs insulated with unfaced or kraft-faced insulation. Overlap the seams and tape them with contractor's tape. Use flexible caulk to seal the polyethylene to abutting surfaces.

in position while you string and staple the portion below it.

3 Installing the Vapor Barrier. A separate vapor barrier is recommended for unfaced batts and those faced with kraft paper. Kraft paper offers only a marginal vapor barrier even when the joints are sealed. The foil on foil-faced insulation can serve as a vapor barrier, but only if extreme care is taken to seal all joints and overlaps with duct tape and/or caulk. To ensure a continuous vapor barrier regardless of the type of insulation, staple a sheet of 6-mil polyethylene over the face of the rafters. Overlap the seams and tape them with contractor's tape. Use tape and/or flexible caulk (such as polybutylene or polyurethane) to seal the polyethylene to abutting surfaces (such as end walls, floors, etc.).

Adding Rigid Foam

If the ceiling to be insulated is too shallow to hold all the necessary blanket and batt insulation, do not despair. By adding rigid foam insulation across the underside of the framing, it may be possible to beef up the total R-value without significantly encroaching into the headroom below.

Begin by completing all rough electrical wiring. Then stuff as much blanket insulation as possible between the rafters, being sure to leave an airspace above for ventilation. Then select a type and thickness of rigid foam (see page 14, "Choosing the Right Insulation") that makes up the rest of the desired R-value.

When choosing a type of rigid foam be sure to consider cost, the R-value of the board, and whether to use foil facing or a separate vapor barrier. The foam is installed between runs of wooden strapping placed across the rafters. Try to match the thickness of the foam to the strapping to be used. Standard thicknesses of strapping are 3/4 inches for 1x3s, and 1½ inches for 2x2s, 2x3s and 2x4s.

1 Marking Lines for Strapping. Measure along one sloped side of the ceiling, marking off intervals of 24 inches. Repeat this process at the other side. Have a helper hold one end of a chalkline while you snap a line across the bottoms of the rafters (or ceiling, if enclosed).

2 Attaching the Strapping. Use 2x3s or rip 2x4s down the center to make two pieces of about the same size as a 2x2. Starting at the top of the ceiling, place a piece of strapping so that its top edge is on the chalk line. Then use 2½-inch-long drywall screws or 20d galvanized box nails to attach it to the rafter. If screws are used, predrill the strapping before inserting the screws.

3 Cutting the Foam. Lay a sheet of rigid foam insulation over a cutting surface, such as scrap plywood or carpet. Measure and mark pieces to fit between the strapping (22½ inches for 2x2 strapping). Using a drywall T-square or other straightedge as a guide, cut through the foam. Use a paring knife to make two or three passes, cleanly cutting through the foam and facing material.

1 Snap chalk lines 24 in. apart horizontally over the exposed edges of the framing or ceiling finish (if enclosed).

2 Use 2½-in. screws or 20d galvanized box nails to attach strapping to the roof framing. If using screws predrill the strapping to make the job easier.

3 Use a paring knife and a drywall T-square (or other straightedge) to cut rigid foam. Place an old piece of carpet or plywood underneath to provide a cutting base.

4 Fit each cut piece between the strapping and secure with duct tape at the corners.

4 Inserting the Foam. Fit each piece of cut foam snugly between the strapping. Cut smaller pieces, as necessary, to trim out the edges. Put pieces of duct tape over the corners and where necessary to hold the cut pieces in place until the ceiling finish material is installed. When all foam is in place, install a polyethylene vapor barrier (see page 35, "Rigid Foam over Gutted Walls") followed by the ceiling finish material.

Blocking Heat with Radiant Barriers

In hot climates, keeping summer heat out is as important as preserving heat during the winter. The most intense heat enters the house through the roof. The best defense against the sun's radiant energy is a radiant barrier placed under the roof. If you live in an area exposed to long, hot summers and your attic has an R-value of less than 30, a radiant barrier may prove to be cost-effective.

1 Preparing the Way. It is necessary to maneuver around the attic, so if the ceiling is not yet insulated, lay down planks over the joists. If the ceiling is insulated, use strips of plywood to spread out your body weight and that of a stepladder.

2 Cutting the Material. Measure the distance between rafters or trusses and cut the radiant barrier into strips of the same width.

3 Attaching the Material to the Roof. Start at a convenient location. Staple the radiant barrier strips (foil facing into the attic) to the underside of the roof sheathing between the framing. Use 1/4-inch staples spaced 6 inches apart along both edges.

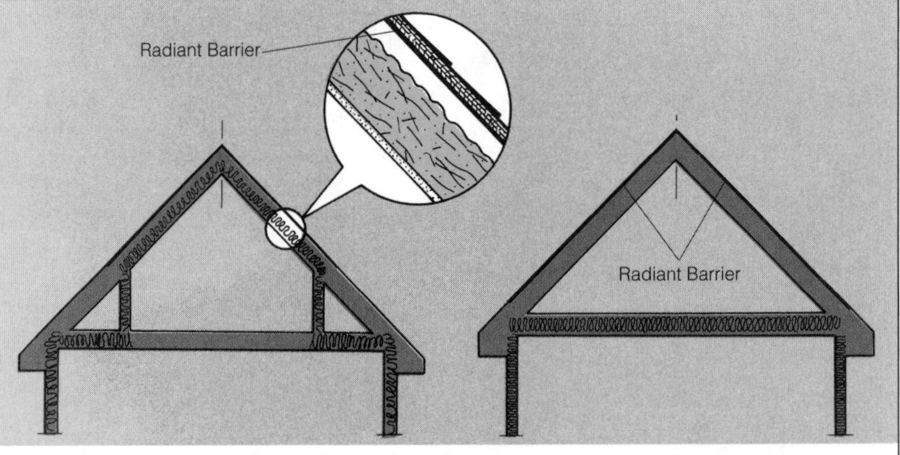

Blocking Heat with Radiant Barriers. Radiant barriers work best when placed on the underside of the roof sheathing. In cathedral ceilings an airspace must be maintained between the radiant barrier and the insulation. In unheated attics insulated at the floor (right) the entire attic serves as the required airspace.

1 Install a portable light and lay down planks over the joists.

2 Measure the distance between rafters or trusses and cut the radiant barrier material into strips of the same width.

3 With the foil facing into the attic, staple the radiant barrier strips to the underside of the roof sheathing.

Basements & Crawl Spaces

If you live in a cold climate, as much as one third of the heat that leaves the house may be escaping through the basement. Much of your basement heat probably seeps directly through the walls into the soil, while some of it may drift out through cracks around the band joist, windows or hatchway.

Reducing heat loss is reason enough to insulate and seal basement walls, but if all or part of the basement is going to be used as an office space, den or entertainment room, you will want to make it as comfortable as possible.

Water & Dampness

Water and dampness plague many basements, so all of the sealing and insulating done to the foundation must include measures to keep out moisture. Controlling basement moisture begins outside. Most of the water that seeps through cracks in the foundation originates at the surface, so the first line of attack is to direct runoff water away from the foundation wall.

If the roof drains directly onto the ground by the affected wall, install a gutter under the eaves. An appropriate gutter will have a downspout that funnels water to a single point on the ground under the eaves. At this point, dig a drainage channel to carry the runoff away from the foundation.

Next, examine the ground near the foundation. If it slopes toward the house, create a slope that directs runoff water away from the wall. If possible, carve the drainage channel deep enough to line it with a sheet of impermeable material (such as 6-mil polyethylene) before covering it with topsoil.

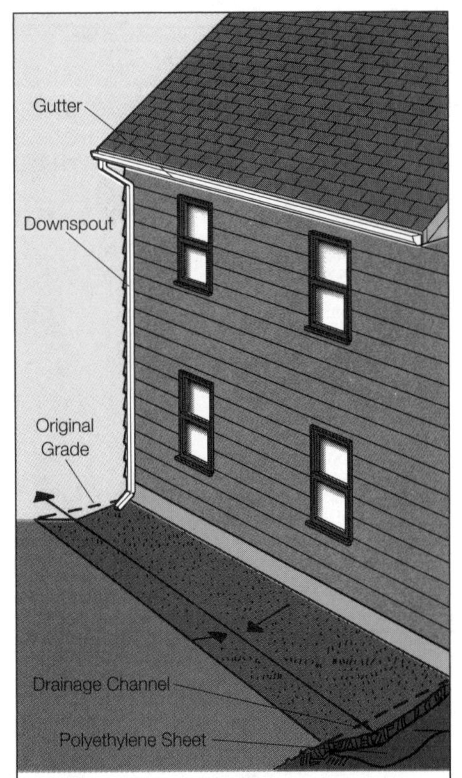

Water & Dampness. To prevent water from leaking through the foundation, install a gutter and downspout on the eaves. Get one that has a pipe or channel to carry the runoff away from the house.

Basement Insulation: Walls or Ceiling?

If the basement does not include living space, it is simpler to insulate the floor between the main floor and the basement (left), than it is to insulate the walls. This approach is reasonable for those living in mild climates, but for cold climates it makes better sense to insulate the basement walls. Here is why:

- Insulating the floor means insulating below every warm air duct and water pipe to keep them warm—this is no small task.
- Furnaces and water heaters give off waste heat anyway, so why not let that heat contribute to the heated part of the house? Otherwise it just goes to waste.
- If all or part of the basement eventually gets finished for living space the walls will have to be insulated anyway .

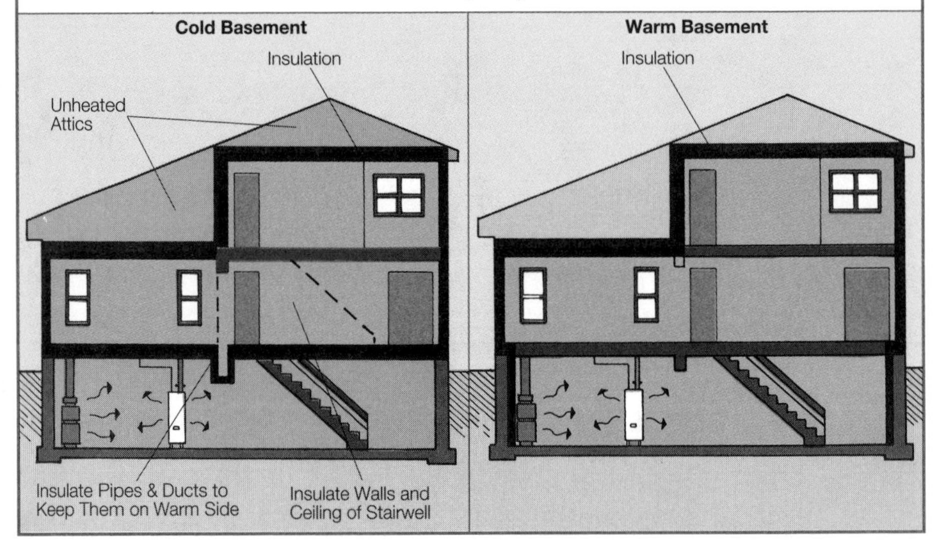

Insulation added to the outside of the foundation can be combined with waterproofing and damp-proofing measures, but this approach is easier with new construction than it is with an existing foundation. Stopping water leaks from inside is a harder job and less likely to be successful. Concrete and concrete block walls have many cracks at the joint lines. These cracks develop as the concrete walls shrink over time. Various materials such as epoxy or cement-based compounds can be injected into the cracks from the inside to control water penetration. This is specialized work, so call in a basement waterproofing expert if the leakage problem has to be solved from inside the house.

The dampness present in basement walls originates from moisture-producing sources found both inside and outside the house. Standing water at low spots on the basement floor, stacks of green firewood, and unvented laundries release moisture into the air. To quickly solve these problems store firewood outside, plug up the source of the standing water, and hook up the clothes dryer to a duct that is vented to the outdoors.

Dampness also can be caused by moisture in the soil which is absorbed through the many tiny capillaries in concrete and masonry walls.

The resulting dampness can be unhealthy to you, as well as unfriendly to objects you want to keep dry. As with waterproofing, the first line of defense against dampness is a good damp-proof barrier located on the outer face of the wall. A vapor barrier placed between the masonry and new wall finish does the trick.

Insulating Foundation Walls

Ideally, insulation is installed on the outside of a foundation wall. This keeps the foundation wall warm, which, in turn, helps keep the basement warmer in winter and cooler in summer. Outside insulation can be combined with waterproofing.

Insulating the outside of a foundation is, in most cases, too big a job for a do-it-yourselfer. The exceptions being new construction where a new foundation has yet to be backfilled. To do the job on an existing house, trenches have to be dug completely around the house—a difficult, often impractical task. Often the only practical solution is to insulate the inside.

Caution: *Because insulating the foundation from the inside surface can result in a frozen foundation, those who live in very cold climates are advised to discuss this job with their local building inspector before proceeding.*

Insulating Basement Walls with Blankets or Batts

The easiest way to insulate the inside of a basement wall is to build a stud wall (made from 2x3s or 2x4s) inside the foundation. Build it as far from the foundation wall as required to achieve the desired thickness of insulation. Because of moisture that is absorbed through the porous masonry or concrete of the foundation, most experts recommend installing a vapor barrier next to the wall before the insulation is applied.

1 Sealing the Band Joist Area. Begin by cutting pieces of 6-mil polyethylene sheet to cover the band joist (the edge around the basement ceiling). Cut pieces to fit between the floor joists and overlap the top, sides and room edge of the mudsill. Cut long pieces for the continuous band joists at the opposite sides. To secure the polyethylene, run a continuous bead of adhesive (such as polyurethane caulk) around the edges. Then press the pieces of polyethylene into the caulk and staple through both using 1/4-inch staples.

2 Extending the Vapor Barrier. Apply a bead of adhesive along the edge of the polyethylene that overlaps the mudsill. Then apply dabs of adhesive down the wall. Staple 6-mil polyethylene sheet to the edge of the mudsill and press it into the dabs of adhesive, overlapping the side walls and floor by an inch or so.

3 Insulating the Band Joist. Cut pieces of blanket or batt insulation to fit snugly into the band joist spaces.

1 Cut pieces of 6-mil polyethylene sheet to fit. Then, press each piece into a continuous bead of polyurethane caulk (or other adhesive) and staple with 1/4-in. staples.

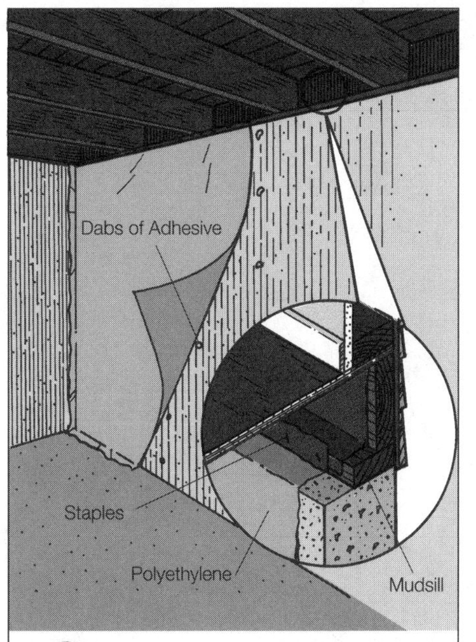

2 Attach the polyethylene with adhesive and staples into the mudsill.

3 Cut pieces of blanket or batt insulation to fit snugly into the band joist spaces between floor joists and install longer pieces at the sides.

4 Attaching the Sill and Top Plates. Attach the sill (bottom) plate to the floor so that the distance from the primary wall to the outside edge of the plate is 3½ inches (for 2x3 framing) or 5½ inches (for 2x4 framing). Lay down two parallel beads of construction adhesive and press the sill into it. Lay bricks or other weights over the sill until the adhesive cures (usually this takes 4 hours, but test it by gently wiggling the piece). Nail a top plate to the ceiling joists at the same distance from the primary wall.

5 Securing the Studs. Cut two 14½-inch-long nailer pieces from the framing lumber and use 12d nails to attach them to the top and sill plates. Then cut a stud to fit between the plates and toenail it into place against the nailers. Repeat the process across the length of the wall until the framing is complete.

6 Installing Batts. Cut 1½-inch-wide strips of insulation and fit them into the spaces between the studs and the foundation wall. Then fill each bay with a length of 15-inch-wide insulation. Push each piece toward the wall, then gently pull out the edges so that it fills the space completely and without gaps. For faced insulation, pull the edge tabs over the face of the studs and staple in place with 1/4-inch staples.

Insulating Basement Walls with Rigid Foam

Another way to insulate the inside of a basement wall is to install rigid foam between strapping. This method may be a bit more difficult than building and insulating inside a secondary stud wall, but the insulated wall encroaches less with this approach and offers an easier way to install a vapor barrier. Unlike blanket insulation, building codes require that foam be covered with a noncombustible material such as 1/2-inch drywall.

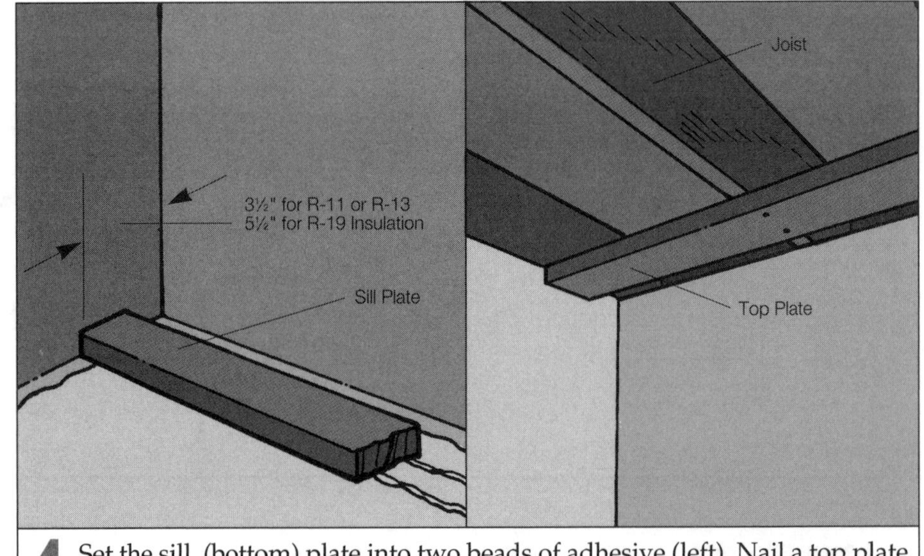

4 Set the sill (bottom) plate into two beads of adhesive (left). Nail a top plate to the ceiling joists at the same distance from the primary wall (right).

5 Cut and nail two nailers to the top and sill plates. Then cut a stud to fit, and toenail it to the nailers. Repeat the process across the length of the wall.

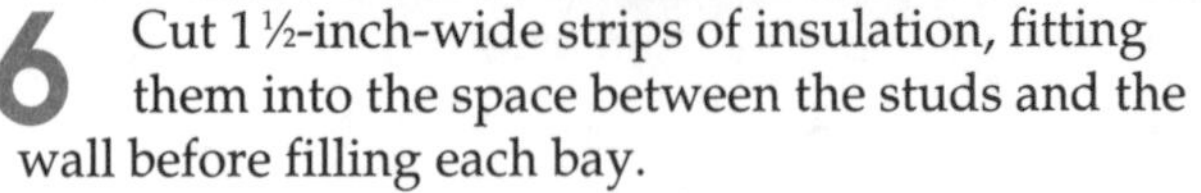

6 Cut 1½-inch-wide strips of insulation, fitting them into the space between the studs and the wall before filling each bay.

1 Attaching Vertical Strapping. Select strapping to match the thickness of the foam to be used. For 1½-inch-thick foam use 2x2 or 2x3 strapping. Begin by running a bead of construction adhesive around the base of the wall, then press a base strapping strip into the adhesive. Mark off 8-foot intervals along the length of the wall. Apply a bead of construction adhesive from the floor to the top of the wall at each location and press a length of strapping into the bead. To keep the verticals from falling over before the adhesive sets, tap a masonry nail into the top of each.

2 Attaching Horizontal Strapping. Starting at the top or the bottom, mark the wall with horizontal lines set 24 inches apart. These marks locate the centers of the strapping and also are in the right position to serve as supports for the drywall finish. Cut pieces of strapping to fit between the verticals and around windows or other objects. Apply beads of adhesive and press strapping into it; then check for level. To support the horizontal pieces until the adhesive sets, either drive a few masonry nails through them, or drive temporary nails into the sides of the verticals and rest the horizontal pieces upon them.

3 Running Electrical Wiring. If the room is going to be finished, electrical outlets will most likely be installed near the floor. Run cable from a junction box in the ceiling, down the sides of the vertical strapping, and into boxes that attach to the strapping with cable staples. Keep the cable 1¼ inches behind the face of the new wall surface (the National Electrical Code requires this distance to prevent surface nails from puncturing the cable). If this is not possible, use a metal plate in places where cable passes through strapping.

4 Cutting the Foam. Lay a sheet of rigid foam insulation over a cutting surface, such as plywood or carpet. Measure and mark pieces to fit between the strapping (22½ inches for 2x2 strapping). Using a drywall T-square or other straightedge as a guide, sever the foam and facing material with a paring knife, making two or three passes before cleanly cutting through it.

1 Install base strapping into the adhesive. Mark off 8-ft.-intervals, apply a bead of construction adhesive, and press a length of strapping into each bead.

2 Install horizontal straping 24 inches on center. Fit strapping around windows and other objects.

3 Run electrical cable from a power source down the sides of the vertical strapping. Attach a metal protective plate in places where the cable comes closer than 1¼ in. to the final finish surface.

4 Mark pieces to fit between the strapping. Lay a sheet of rigid foam insulation over a cutting surface and use a paring knife to cut through the foam. Use a drywall T-square or other straightedge as a guide.

5 Insert the panels between the strapping, holding them temporarily in place with tabs of duct tape.

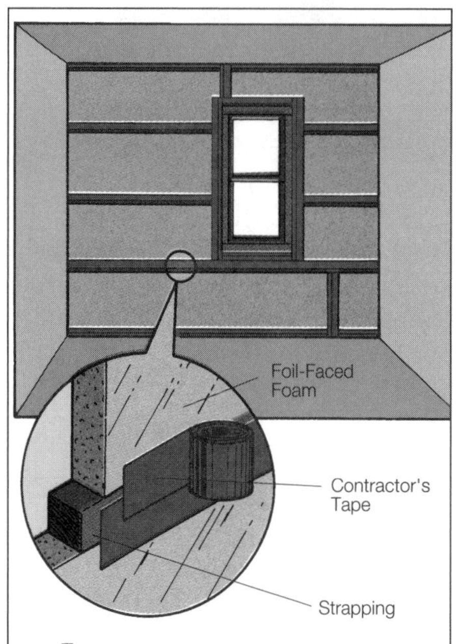

6 Use contractor's tape or duct tape to close the joints of foil-faced foam. Use a compatible caulk to seal abutting joints of walls, floors and ceilings.

5 Inserting the Panels. Insert the panels. Place tabs of duct tape at the joints, corners and midpoints to hold the foam panels in place until the next step. Cut grooves in the backside of panels to accommodate electrical cable.

6 Taping the Joints. If unfaced foam was used, apply a continuous layer of 4- or 6-mil polyethylene vapor barrier over the foam and strapping. For foil-faced foam, simply use contractor's tape or duct tape to tape all joints between foam and strapping. Make two or more passes to ensure continuous coverage. Use a compatible caulk to seal the abutting joints of walls, floors and ceilings.

Insulating the Basement Ceiling

If the basement ceiling will be insulated instead of the walls, wrap all water pipes and air ducts to isolate them from the cold.

Wrapping Ducts

1 Enclosing the Sides and Bottoms. Cut lengths of kraft- or foil-faced blanket insulation long enough to wrap around the sides and bottom of the duct with enough spare to overlap the joists. Do not extend or tuck the insulation over the top of the duct because it will isolate the duct from the above heated space. Cut the insulation away where it overlaps the joists and staple the paper or foil to the joists. Then use duct tape to wrap the seams.

2 Closing off Duct Ends. Leave flaps of insulation overhanigng the duct ends. Fold the long flaps over the end of the duct and tape them together. Then fold the stubs at the bottom into the duct bottom and tape.

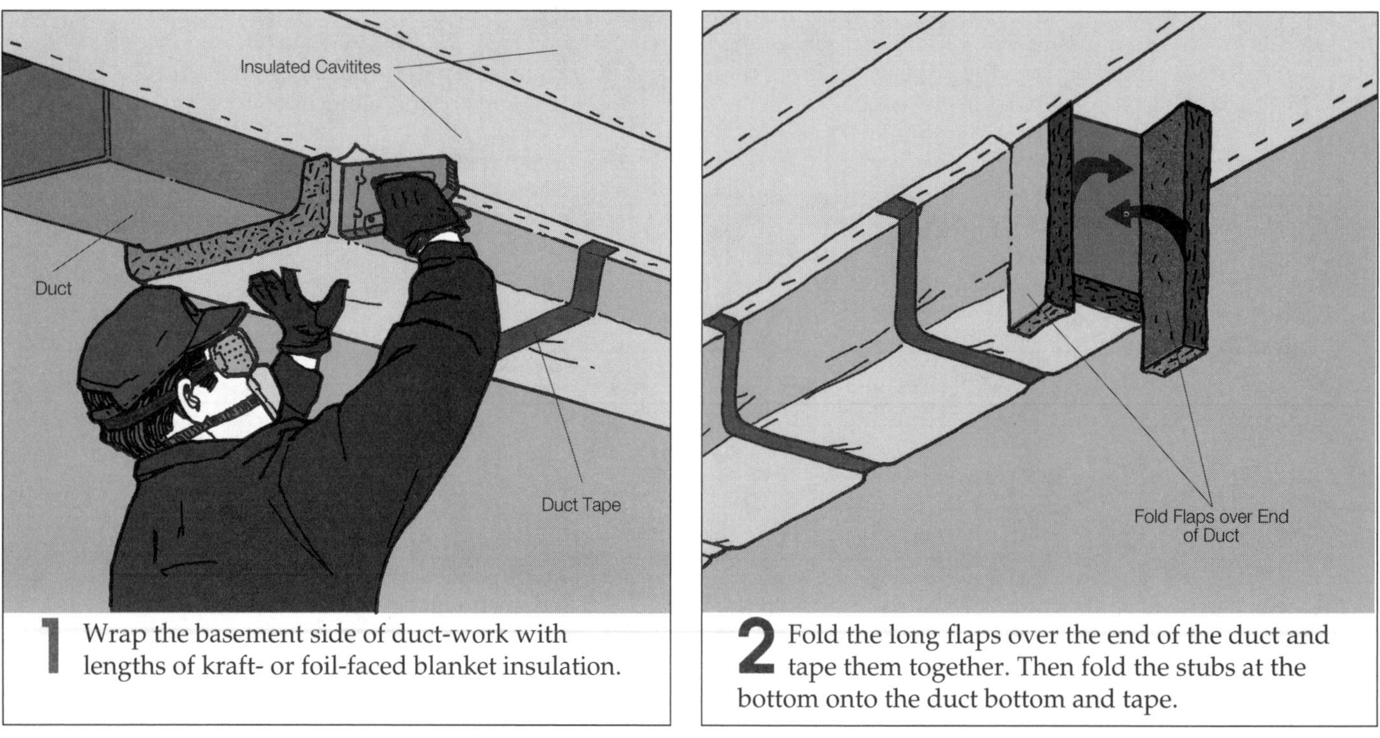

1 Wrap the basement side of duct-work with lengths of kraft- or foil-faced blanket insulation.

2 Fold the long flaps over the end of the duct and tape them together. Then fold the stubs at the bottom onto the duct bottom and tape.

Insulating Water Heaters and Piping

An insulated basement ceiling results in an unheated basement. A furnace or boiler that is located in the basement may give off enough heat to prevent water pipes from freezing, but relying on it is a gamble. The heat may be turned down or off completely if the house is vacated for some reason. To minimize the risk of freezing pipes, wrap all water-carrying pipes whether they carry hot or cold water. To save additional energy, wrap the water heater as well.

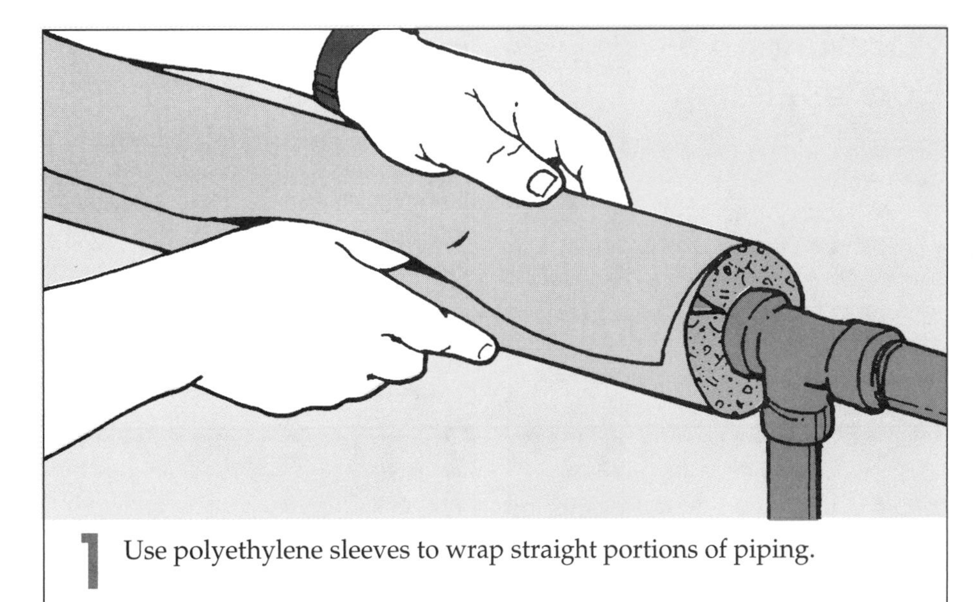

1 Use polyethylene sleeves to wrap straight portions of piping.

1 Attaching Pipe Sleeves. Use polyethylene sleeves to wrap long portions of water pipes. Slip the sleeves over the pipe through the precut slit. Then close the attached lap to seal the joint. If the sleeve does not have an attached lap, wrap it with duct tape every 24 inches or so to keep the slit closed.

2 Covering the Joints and Bends. Insulate turns and bends between sleeves by wrapping pipe wrap (narrow rolls of blanket insulation) spirally around pipes, maintaining a half-width overlap.

3 Wrapping a Water Heater. The easiest way to insulate a water heater is with a ready-made kit obtained from a hardware or home center store. Faced insulation can be used instead if you have it. Wrap lengths around the heater horizontally and use duct tape to tape the joints. Leave all controls and valves exposed. If the water heater is fueled by oil or gas, cut the insulation short of the vent stack by at least 2 inches and do not insulate the vent. Also, be sure to keep insulation away from the air intake.

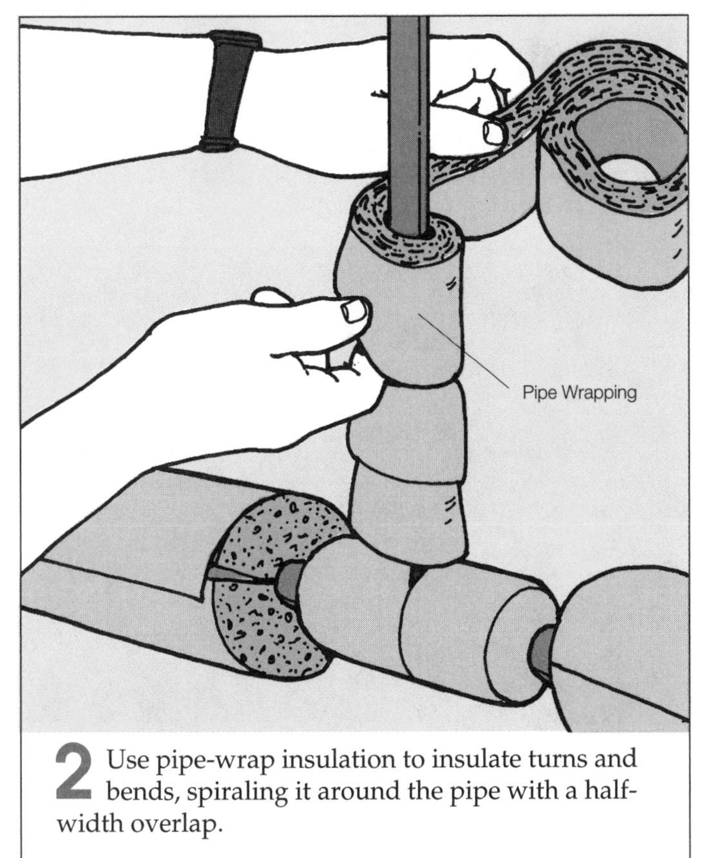

2 Use pipe-wrap insulation to insulate turns and bends, spiraling it around the pipe with a half-width overlap.

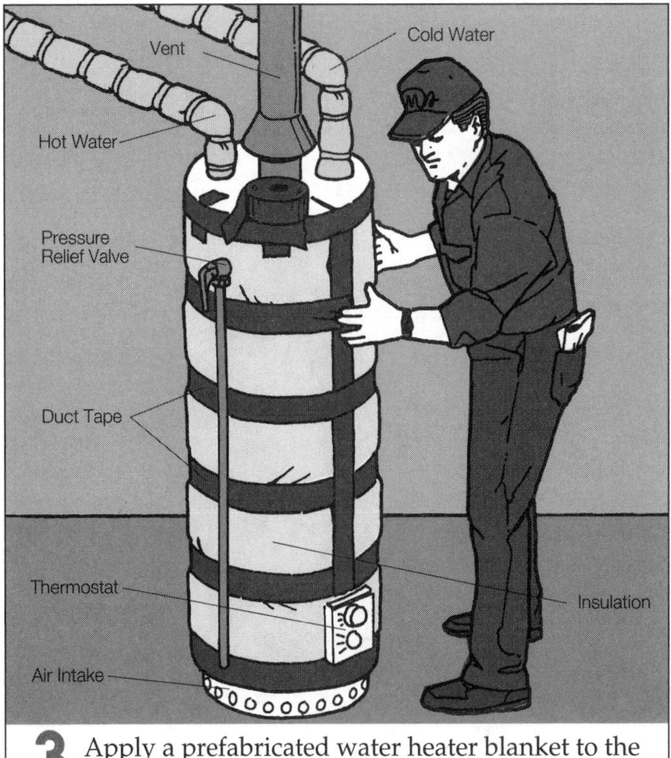

3 Apply a prefabricated water heater blanket to the unit or wrap lengths of faced blanket insulation around the heater horizontally. Tape the joints with duct tape.

Warming up a Concrete Floor

During the winter, people feel more comfortable when there is a warm floor underfoot. A concrete slab quickly drains heat from the interior to the cold ground below, chilling the air above it in the process. Before insulating the top of the slab, consider how the space above it will be used. Slabs that are situated several feet down, such as in a basement, are not as cold as slabs near the surface, such as in a garage. If the basement will remain unfinished or finished for occasional uses only, such as a place to put the pool table, it probably does not make sense to insulate the floor. However, if people will be sitting on the floor in a family room, you will want it to be warm. Floor insulation also is a good idea for those who want to convert part of the garage into an office.

Concrete floor slabs are insulated with rigid foam that is set between wood supports called sleepers. The method of installation is very similar to the way foam is installed on walls between strapping. Select dry, straight 2x4s to use for sleepers. Top the sleepers and foam with a plywood subfloor, and finish the floor with tiles, resilient flooring or carpet.

Installing an Insulated Subfloor

1 Putting Down a Vapor Barrier. After sweeping the floor slab clean of dust and debris, cover it with 6-mil polyethylene, overlapping each seam by a minimum of 6 inches. Lift up the edges of the polyethylene and use a caulking gun to put down dabs of construction adhesive to hold it in place.

2 Installing Perimeter Sleepers. Nail sleepers around the perimeter of the room using 2¼-inch-long masonry nails. If the lumber is dry and straight, a nail or two every several feet should suffice. Then mark these plates for additional sleepers 24 inches on center. This spacing will be suitable for 3/4-inch plywood subflooring.

Caution: *Never strike masonry nails with a standard hammer. The nails are likely to damage the hammer head, which could propel metal shards at you. Instead, use a hand drilling hammer which has a head specially designed for striking concrete objects. Still, be sure to wear eye protection.*

3 Installing Interior Sleepers. Align the interior sleepers square to the marks on the perimeter plates. Use one nail at the end of each board and one about every four feet. Butt sleepers end-to-end if they do not reach completely across the room.

1 Roll polyethylene over the floor, unfold it, and overlap the seams by at least 6 in. Press the polyethylene into dabs of construction adhesive.

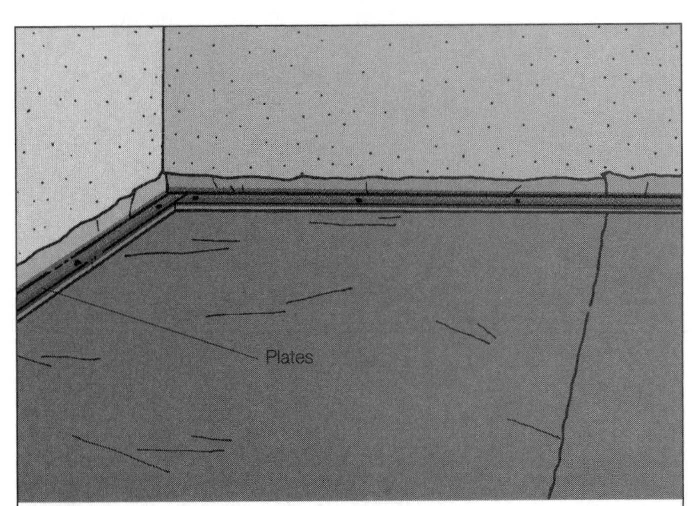

2 Nail sleepers around the perimeter of the room, butting joints as needed. The pieces should be straight and flat. Secure them to the slab with with masonry nails spaced every few feet.

3 The "joists" of this subfloor are 2x4 sleepers spaced 24 inches on center. Align them to provide solid bearing for the plywood, then nail them to the floor.

4 Cut rigid foam to fit between the sleepers and drop it in place.

5 Nail plywood to the sleepers with 6d nails.

4 Inserting Foam. Medium-density extruded polystyrene foam is best for concrete floor slabs. Use a thickness that matches the thickness of the sleepers. Cut each piece to fit between the sleepers and insert it in place.

5 Attaching the subfloor. Use 3/4-inch-plywood subflooring, either straight-edges or tongue-and-groove. Cut the plywood to span across, rather than parallel to the sleepers. Do this in a staggered pattern. Attach the plywood to the sleepers with 6d nails. Place the nails every 6 inches around the perimeter and 12 inches over intermediate supports.

Insulating Crawl Spaces

Crawl spaces often are dark, dirty places filled with cobwebs and sometimes rodents. Still, they are an important part of the insulating plan. An uninsulated crawl space is a source of moisture from the ground and a major source of heat loss, especially in locations that endure cold winters.

As with full basements, crawl spaces can be treated as cold areas outside of the insulated boundary of the house, or as warm areas within the boundary. Isolating them outside the heated boundary requires insulating the floor. This technique has all the drawbacks mentioned in connection with basements. Including the crawl space within the heated boundary is easier to do and it is more effective in cold climates. Some local codes require the use of outside vents placed in the crawl space wall to remove moisture. If vents are required, make sure they can be closed off when the heat is on to avoid blowing away gains made in insulating the walls.

Creating a Warm Crawl Space

A crawl space can be insulated using the same game plan used for a full basement: insulate the floor or insulate the foundation wall. Insulating the floor creates a cold crawl space suitable for areas that have mild winters. In areas subject to cold winters, however, insulating the foundation walls so that the crawl space stays warm usually is a better choice. If this method is used, make sure that vents in the wall can be closed in the winter.

No matter how the crawl space is insulated, a vapor barrier next to the foundation wall and over the soil is a must to keep moisture from migrating from the soil into the space. With no barrier, moisture can enter and condense on parts of the wood framing, promoting the growth of wood-destroying organisms. It also can find its way into the outside walls and cause paint to peel.

To help make working under the crawl space less troublesome, hang a portable lamp for light. Protect your head and eyes by wearing a hard hat and goggles. Wear a dust mask or respirator.

1 Attaching the Vapor Barrier to the Wall. Unroll the polyethylene sheet on the floor and use 1/4-inch staples spaced at 8-inch intervals to attach it to the mudsill. Run each strip down the inside of the foundation wall and 12 inches over the ground soil. Overlap the seams found at each side by at least 12 inches. If you do not intend to insulate the walls, install the floor vapor barrier next, as described in Step 5 on page 66.

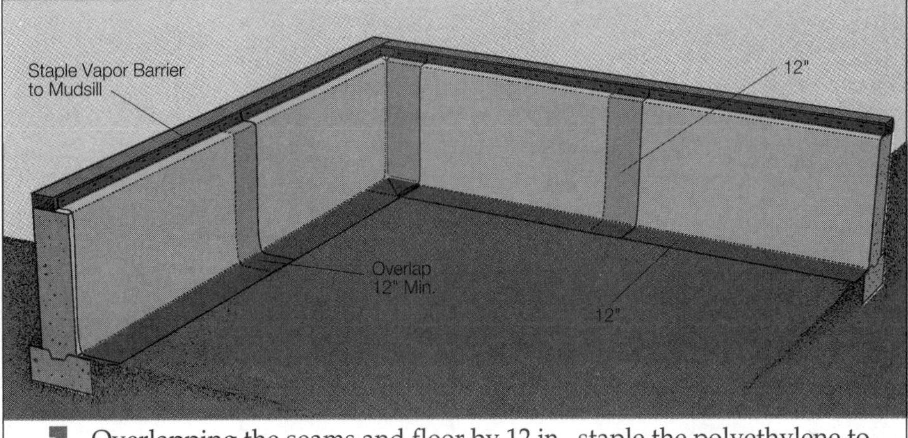

1 Overlapping the seams and floor by 12 in., staple the polyethylene to the mudsill and drape it down the inside of the foundation.

2 Hanging Insulation Parallel to Joists. Measure the vertical distance between the underside of the floor above and the ground, and add 3 feet. Cut strips of kraft-faced insulation to this length. Place one end of each strip next to the band (or rim) joist and tightly against the underside of the floor. Use a 1x3 cleat to secure it to the band joists.

3 Hanging Insulation at Joists Intersections. Cut notches in the ends of the batts to fit around the joists at the walls where joists run into the band joists. Then hang strips down the wall, anchoring them to the edge of the mudsill with 1x3 cleats.

4 Connecting Tabs. Connect each strip of insulation to the next by folding out the edge tabs and stapling them together with an office stapler at 8-inch intervals.

5 Installing a Vapor Barrier over the Soil. When all wall insulation and vapor barrier sheets are in place, cover the soil completely with polyethylene, overlapping the side seams by at least 12 inches. Allow 6 inches or so excess where the ground sheet meets the wall.

6 Securing the Bottom Corners. Place lengths of 2x4 over the top of the insulation and vapor barrier at the joint between the wall and earth to hold the bottom edges in place.

2 Begin hanging batts at the band joist along the wall parallel to the joists.

3 Cut the ends where necessary to fit around the joists at the intersecting walls.

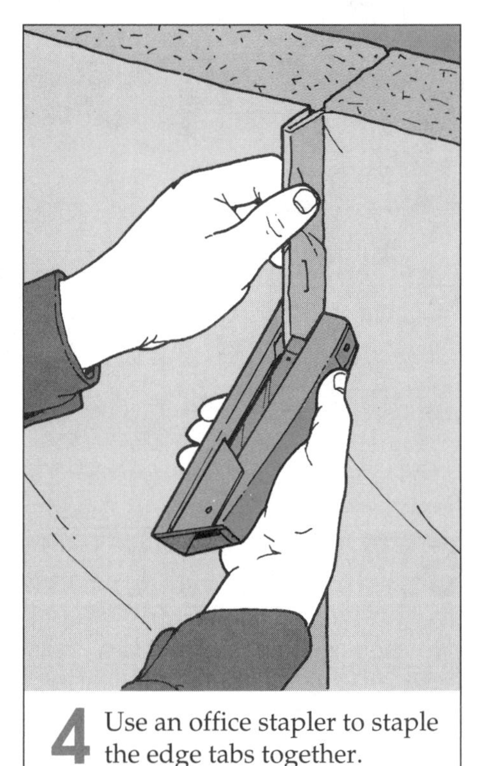

4 Use an office stapler to staple the edge tabs together.

5 Cover the soil with polyethylene sheet. Lift up the batts and tuck the polyethylene over the overlapping edge of the polyethylene hanging down the wall. Overlap side seams on the floor by 12 in.

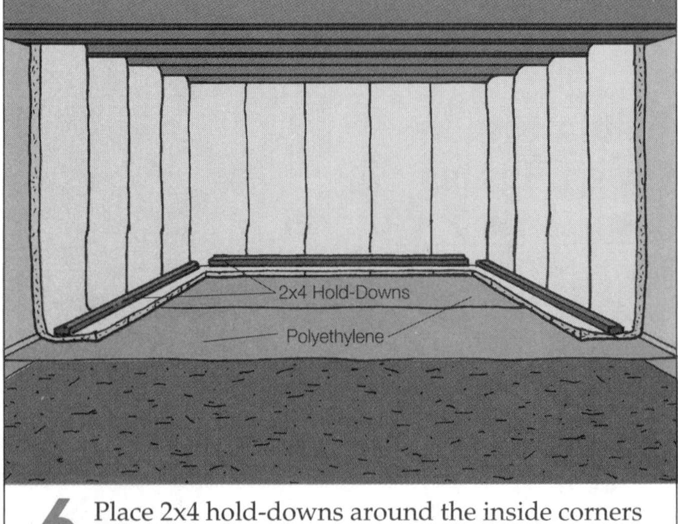

6 Place 2x4 hold-downs around the inside corners of the hanging insulation and vapor barrier.

VENTILATION

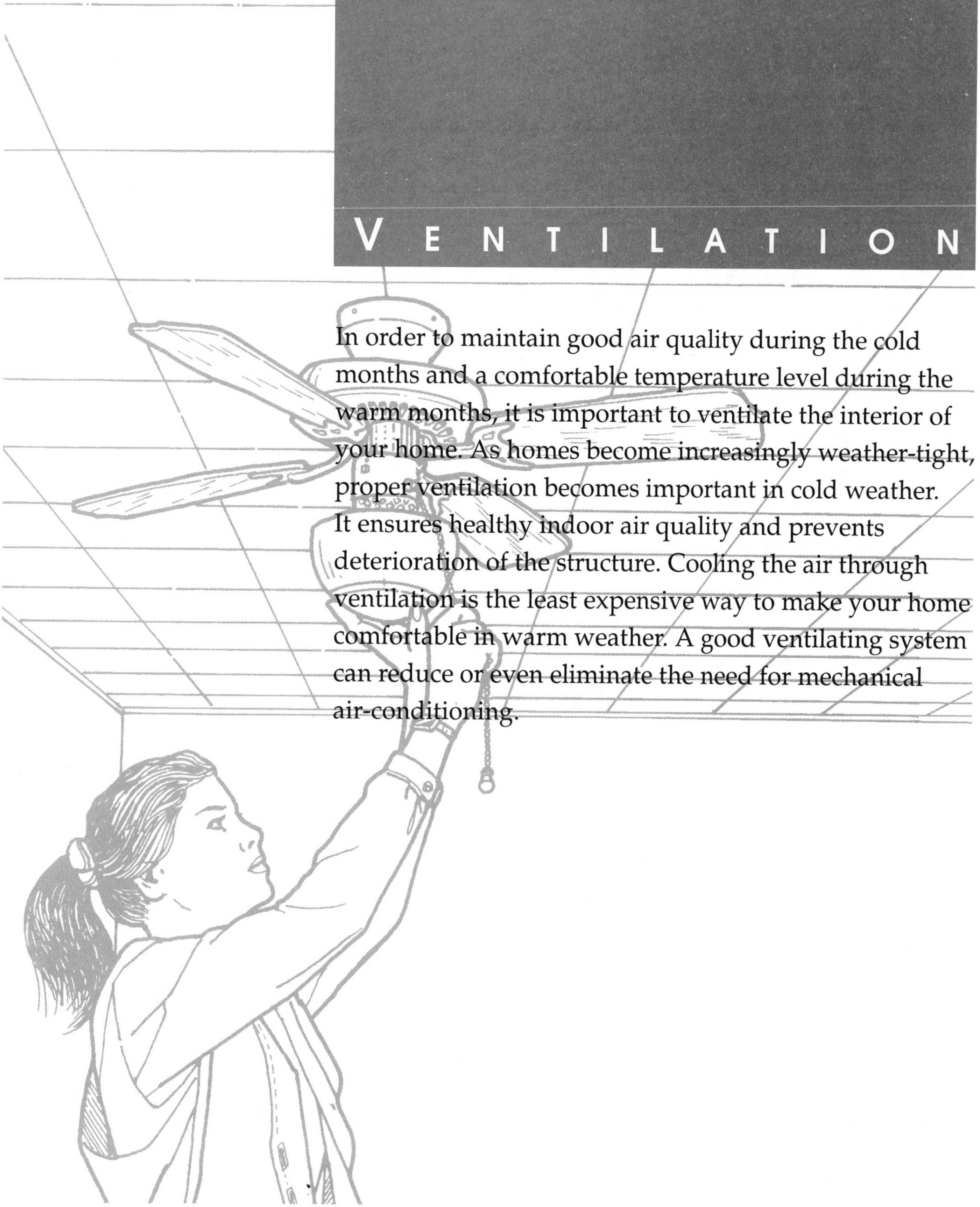

In order to maintain good air quality during the cold months and a comfortable temperature level during the warm months, it is important to ventilate the interior of your home. As homes become increasingly weather-tight, proper ventilation becomes important in cold weather. It ensures healthy indoor air quality and prevents deterioration of the structure. Cooling the air through ventilation is the least expensive way to make your home comfortable in warm weather. A good ventilating system can reduce or even eliminate the need for mechanical air-conditioning.

Ventilating to Maintain Air Quality

In mild weather, obtaining fresh air is as easy as opening a window. The problem is ensuring the supply of fresh air in cold weather. Before people started tightening up their homes to conserve energy, enough air leaked in and out of the many cracks in the walls and roof to maintain a fresh supply of air inside the house. With these unplanned openings sealed off, humidity and contaminants can build up to levels that threaten your health, as well as the health of the structure in which you live. The answer is not to leave cracks unsealed around the windows and doors, but to provide controlled ventilation when the windows are closed.

Choosing an Exhaust Fan

In just one week of winter, a family of four adds about 18 gallons of water vapor to the inside air through breathing, cooking, bathing, and washing clothes and dishes. The best way to control moisture—and the odors it harbors—is to remove it at the source and exhaust it directly to the outdoors. Make sure the clothes dryer exhausts to the outside, rather than into an attic or basement space. Provide an exhaust fan to the outside from the kitchen and bathrooms. Kitchen exhaust fans can be switched on and off manually. A bath fan can be wired to its own switch or hooked up to a light switch.

An effective bath exhaust fan changes all of the air in a room at least eight times per hour, or once every 7.5 minutes. To get the required fan capacity in cubic feet per minute (cfm) divide the volume of the room by 7.5. If the room measures 8 by 9 feet and has an 8-foot ceiling, its volume is 8x9x8 = 576 cubic feet. The fan capacity needed is determined by dividing 576 by 7.5, which equals 76.8 cfm.

Another thing to look into when selecting a fan is its noise level. Fan noise is rated in sones. The quietest kitchen and bath exhaust fans are rated at approximately one sone, about the same loudness as a refrigerator. Louder fans rate up to four sones.

The capacity of kitchen exhaust fans varies from approximately 160 cfm to 600 cfm. Estimating the size needed is not a precise science, and most models have a variable switch that allows the fan to be set anyway. If the duct-work connecting the fan to the outside is very long (over 6 feet), select a fan from one of the higher ranges.

The best place for a bath exhaust fan is in the ceiling or high in a wall above a toilet, tub or shower. A kitchen exhaust fan is most effective in a range hood that is mounted above the cooking area.

Installing a Bathroom Exhaust Fan

There are three common ways to install a bath fan. The easiest is to install the fan in an outside wall so it exhausts without ducting. The other options are to install the fan in the bathroom ceiling and duct it either through the roof or soffit.

Ducting Through an Outside Wall

If the bathroom abuts an outside wall that has a vacant area near the ceiling, it may be possible to mount the fan through a hole in the wall.

When a location has been chosen some detective work is necessary to minimize the possibility that a wire or pipe exists in the wall in that area. Take a look at switches, outlets and lighting fixtures in the bathroom. There is no way to accurately discern where an electrician may have run wires, but if the fan is to be located directly in the shortest path between two electrical devices, it is a good idea to explore further. Turn off the power and pull the devices out of their boxes. Then check the wires for the direction from which they originate.

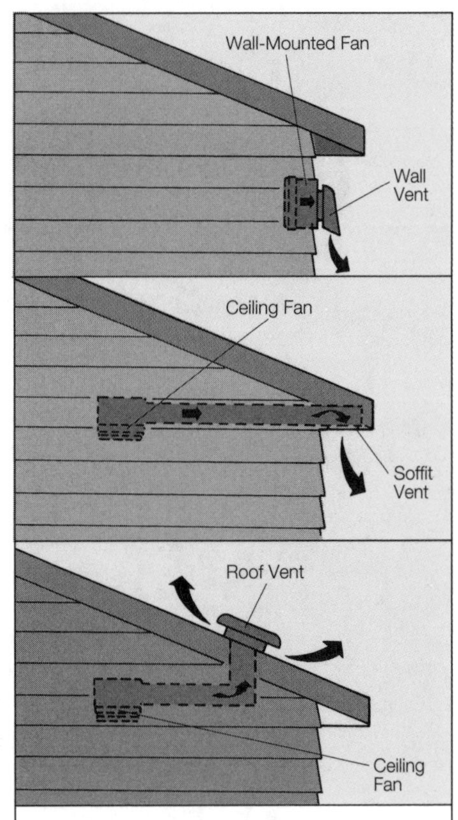

Installing a Bathroom Exhaust Fan. **The easiest and most direct place to mount a fan is in an outside wall. Fans also can be mounted in the ceiling and ducted to a terminal in the soffit or roof.**

Take an educated guess at where the plumbing is located. Those living in a cold climate probably are safe because smart plumbers avoid putting water supply lines in outside walls. However, there are no guarantees. Is there a sink, toilet, shower or bathtub directly above or below the place in which you want to cut a hole? Go into the attic and look for plumbing vent stacks that come through the attic floor.

1 Locating the Hole. Shut off electrical power to the area just in case you hit a wire when cutting through the wall. Drill a series of small holes across the portion of the wall or use a stud locator to locate a clear space between studs. Use a long bit to drill a starter hole completely through the wall. Inspect where the drill comes through the outside wall to see if there is adequate clearance for the outer

1 Locate a clear space between studs. Use a long bit to drill a starter hole completely through the wall.

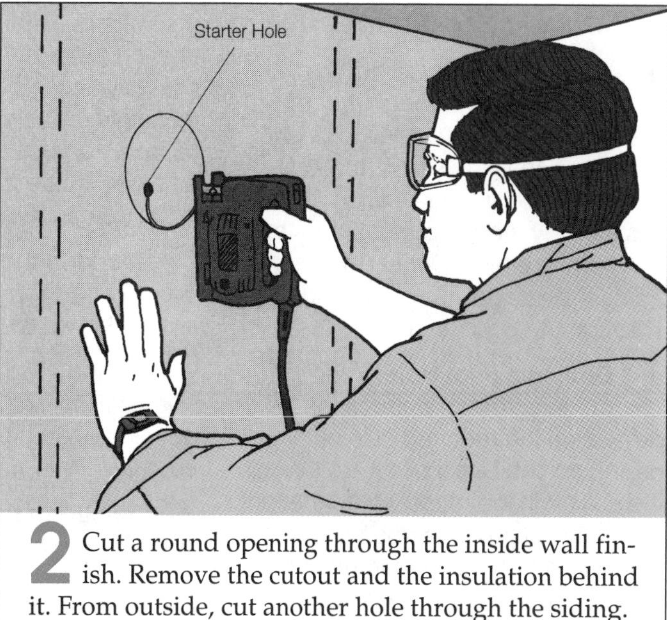

2 Cut a round opening through the inside wall finish. Remove the cutout and the insulation behind it. From outside, cut another hole through the siding.

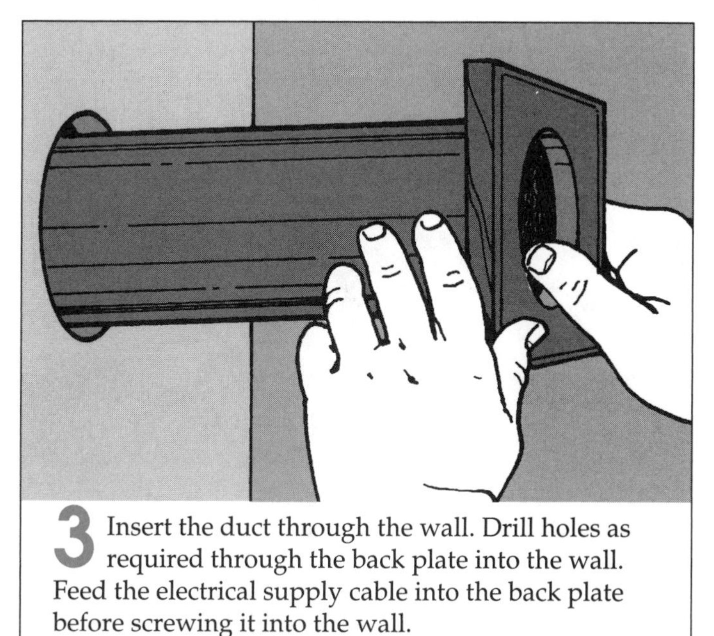

3 Insert the duct through the wall. Drill holes as required through the back plate into the wall. Feed the electrical supply cable into the back plate before screwing it into the wall.

4 Attach the vent louver to the duct and screw it to the wall. If working with clapboard siding, it may be necessary to insert pieces of clapboard behind the plate, to create a flush mounting surface.

vent. Make sure, for example, the outside trim of the fan won't encroach on window casing or other trim.

2 Cutting the Hole. Hold the fan duct against the wall and mark the opening on the inside wall. Insert the blade of a saber saw or keyhole saw into the starter hole and cut a round opening through the inside wall finish. Remove the cutout and the insulation behind it. If there are no pipes or wires obstructing the opening, cut a similar hole through the siding from the outside.

3 Inserting the Duct. Follow the manufacturer's instructions for the correct method of fitting the inside back plate to the wall duct. Insert the duct through the wall. Drill holes as required through the back plate into the wall, then feed the electrical supply cable into the back plate before screwing it to the wall. Screw on the inside plate before installing the cover.

4 Attaching the Outer Louver. From outside, use a screwdriver to stuff bits of fiberglass insulation around the duct. Then attach the vent louver to the duct and screw it to the wall. If working with clapboard siding, it may be necessary to create a flush mounting surface. One way is to cut scraps of siding and nail them upside down to the wall around the opening. After mounting the louver, caulk around the perimeter to seal it to the wall. Finally, complete the wiring as described on page 73, "Wiring a Bathroom Fan."

Installing a Fan in the Ceiling

Whether the fan is to be ducted through a soffit or through the roof, the procedure for installing the fan in the ceiling is the same. If you are planning a ceiling installation, follow this step-by-step sequence and then move on to either "Ducting Through a Soffit" on page 71 or "Ducting Through the Roof" on page 72.

1 Drilling a Pilot Hole. Read the manufacturer's instructions to determine the required size of the ceiling cutout. Drill a pilot hole through the ceiling in the place where the fan will be installed, then push a length of wire through the hole to mark this spot. Bend or tape the wire to hold it temporarily in place.

2 Locating the Opening. Locate the wire in the attic. Measure the distance between the wire and the nearest joist and measure the width of the space between joists. Transfer these dimensions to the ceiling below. Measure across the ceiling from below to locate the midpoint. This is the center of the cutout.

3 Marking the Ceiling. Center the fan housing over the midpoint line and mark the area to be cut.

4 Cutting the Opening. Drill a starter hole at one corner of the opening, then use a keyhole saw to cut the opening.

5 Securing the Fan Housing. From the attic, nail the extension support brackets that come attached to the fan housing to the joists or cut lengths of 2x4 to span between two joists as a support.

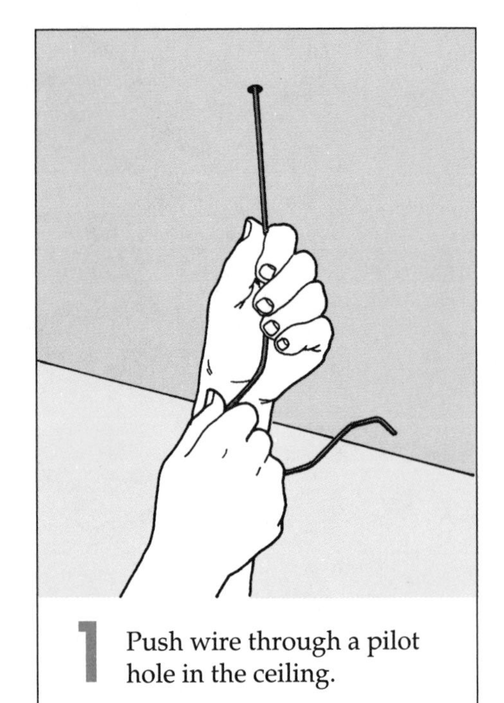

1 Push wire through a pilot hole in the ceiling.

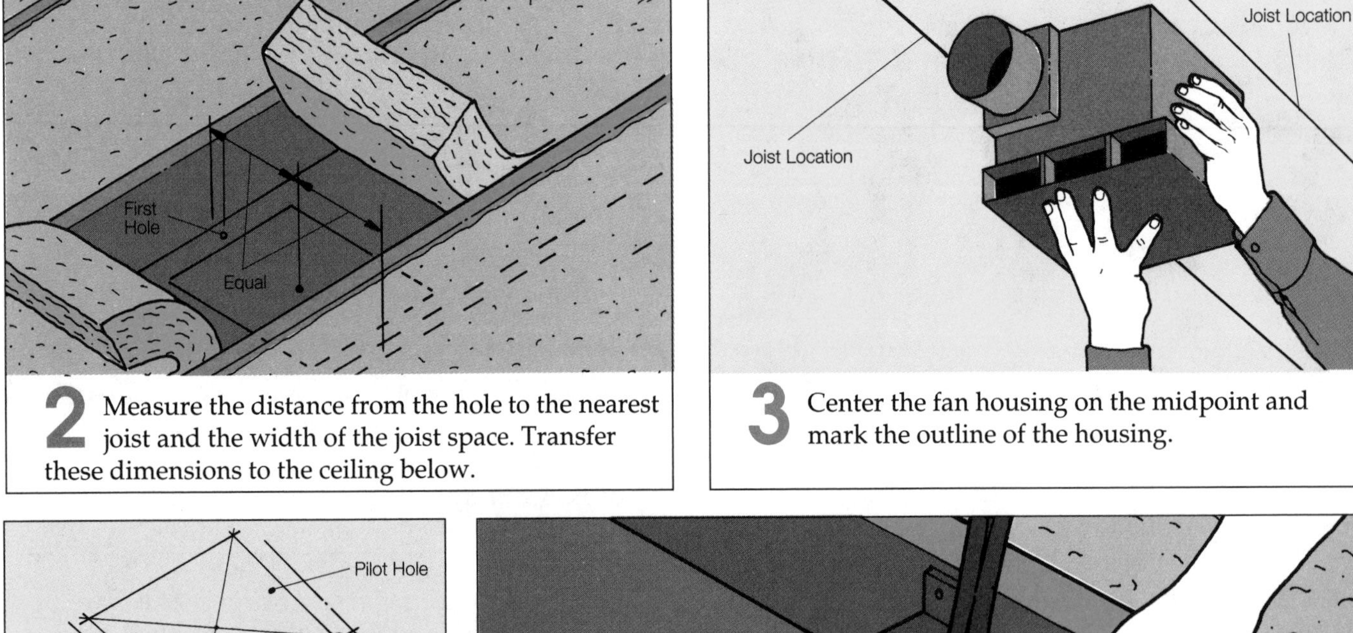

2 Measure the distance from the hole to the nearest joist and the width of the joist space. Transfer these dimensions to the ceiling below.

3 Center the fan housing on the midpoint and mark the outline of the housing.

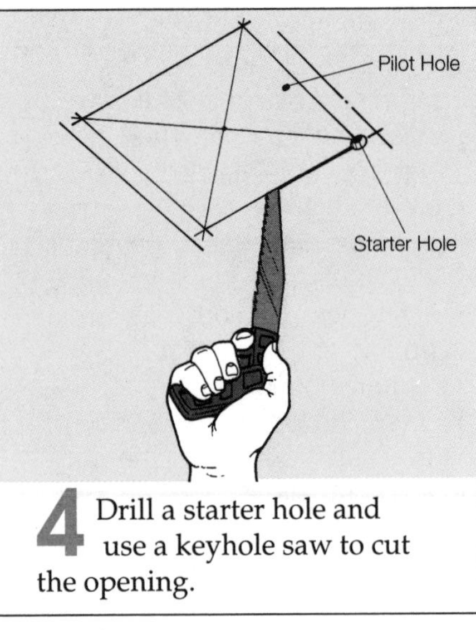

4 Drill a starter hole and use a keyhole saw to cut the opening.

5 Nail the support brackets to the joists.

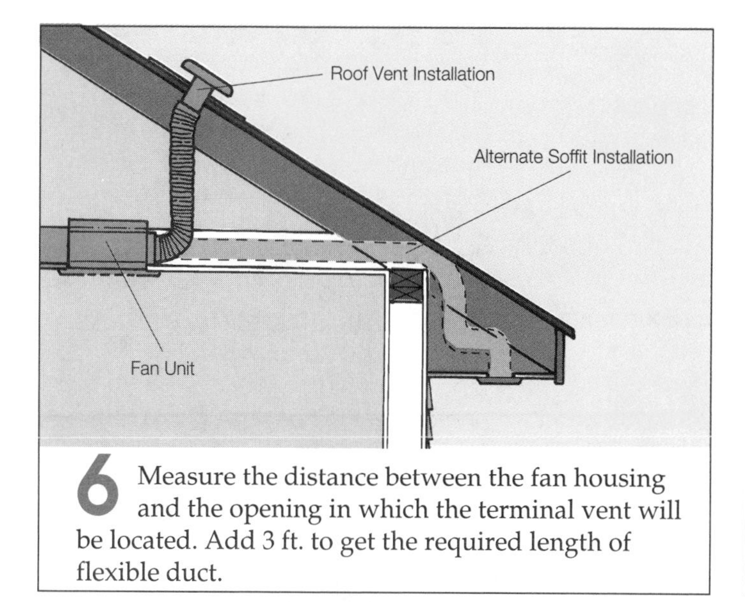

6 Measure the distance between the fan housing and the opening in which the terminal vent will be located. Add 3 ft. to get the required length of flexible duct.

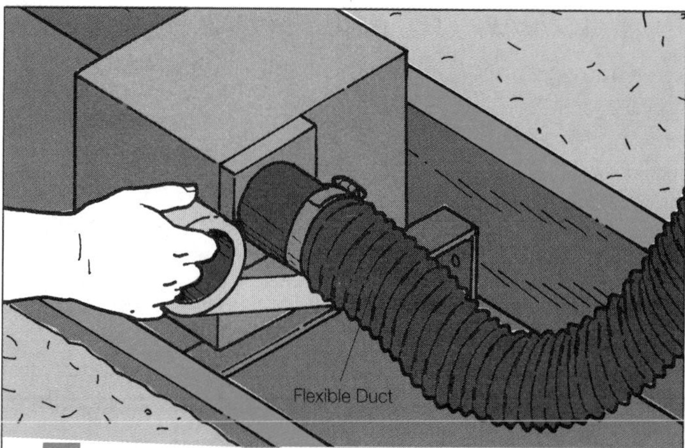

7 Use a hose clamp to attach one end of the flexible duct to the fan outlet. Wrap it tightly with duct tape.

6 Cutting a Length of Flexible Duct. In the attic, measure the distance between the fan housing and the roof (or wall or soffit) opening where the vent will be installed. This distance plus an extra 3 feet is the amount of flexible aluminum duct needed. Match the diameter (usually 4 inches) to the fan.

7 Connecting the Duct to the Fan Housing. Clamp the sleeve of the flexible duct to the discharge opening of the fan housing. Wrap duct tape around the clamp, making an airtight joint.

Ducting Through a Soffit

1 Cutting an Opening in the Soffit. From outside, drill a pilot hole through the soffit in the place where the exit vent will be located. Using the sleeve of the vent cap as a template mark the cutout. Then cut the opening with a saber saw or keyhole saw.

2 Installing a Soffit Vent Cap. Lift up the insulation in the attic ceiling as necessary to extend the duct through the hole in the soffit. From outside, use a hose clamp and duct tape to connect the duct to the sleeve of the vent. Push the sleeve into the hole and screw the vent into the soffit. Restore insulation that earlier was displaced in the attic.

1 Use a saber saw to cut a round opening through the soffit or roof. This opening matches the diameter of the duct.

2 Extend the duct from the fan to the soffit hole. Connect the duct to the sleeve of the vent with a hose clamp and duct tape. Push the sleeve into the hole and screw the vent to the soffit.

Ducting Through the Roof

1 Cutting an Opening in the Roof. Locate the hole in the roof so that the duct butts against a rafter as it passes through the roof. If there is insulation in the rafters, remove it from the area where the duct will be installed. Drive a nail through the roof sheathing to locate the center of the duct hole. Climb up on the roof and locate the nail. Then poke it back down so it is out of your way. Use the sleeve of the vent cap as a template to lay out the hole in the roof. Use a saber saw to cut away the shingles and the roof sheathing.

2 Installing the Duct Extension Clamp. The flexible duct must have a rigid extension to pass through the roof and fit into a hole in the bottom of the vent cap. To hold the extension in place, screw a hose clamp to the side of the rafter through which the extension will pass.

3 Installing the Duct Extension. Use duct tape to attach a 10-inch length of rigid galvanized steel duct to the discharge end of the flexible duct.

4 Installing the Vent Cap on the Roof. Push the duct extension through the hose clamp so that it extends 1/2 inch through the opening in the roof. Tighten the hose clamp. The vent cap has a round hole in the bottom that fits over the duct extension. Cut away shingles and remove roofing nails as necessary so that when the cap is installed, its flange will be covered at the top and sides but not at the bottom. Coat the underside of the roof-cap flange with roofing cement and press it into place. Use more roofing cement to coat the undersides of those roof shingles that will cover the flange. Seal around the edges of the shingles with asphalt caulk.

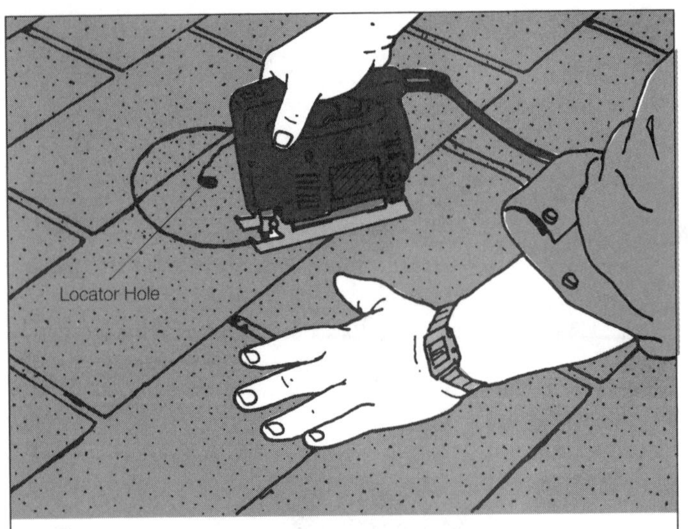

1 Use a nail, driven through the roof from inside, to determine the location of the hole. Use a saber saw to cut out the hole.

2 Hold the extension in place by attaching a hose clamp to the rafter.

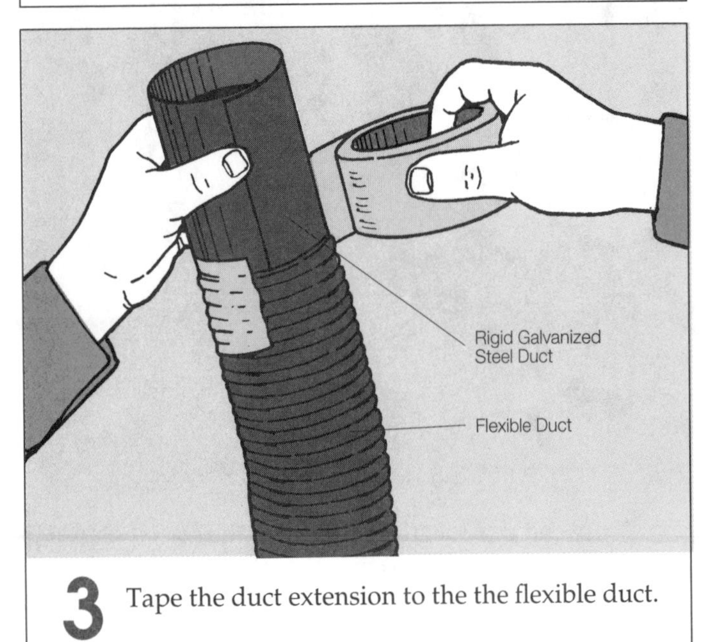

3 Tape the duct extension to the the flexible duct.

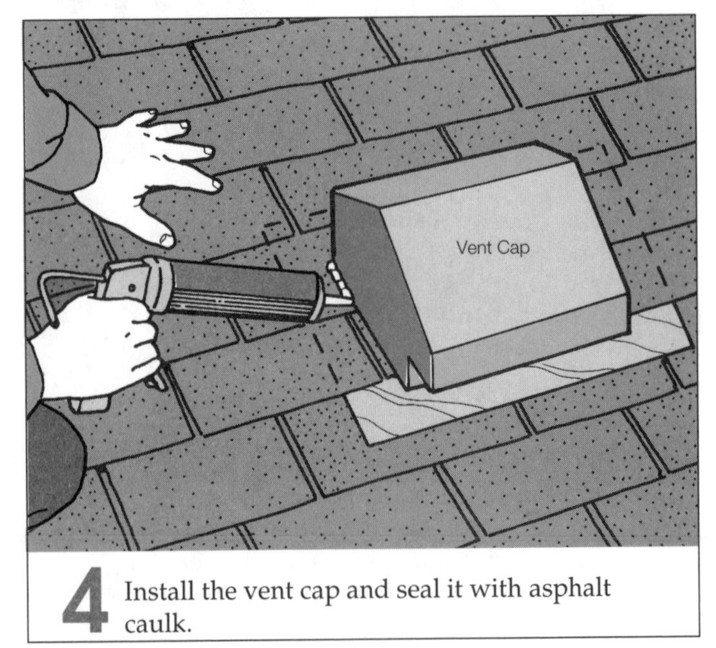

4 Install the vent cap and seal it with asphalt caulk.

Wiring a Bathroom Fan

The electrical part of this project consists of bringing a power cable to the fan motor and hooking it to a switch. If the power source is closer to the fan than it is to the switch, a switch loop is used. If not, the switch can be wired in-line. Both methods are described below.

Wiring a Switch Loop

Use a switch loop if the power source is nearer the fan than the switch. A switch loop brings a power cable into the fan's electrical box, from which another cable, or loop, runs to the switch.

Caution: *Before beginning work go to the electrical service panel and shut off the circuit breaker to the circuit upon which work is being done.*

1 Wiring the Fan. Run two cables into the fan housing; one from the power source and one for the switch loop. Coming from the fan housing itself will be a white wire and a black wire. The housing has a green grounding screw. Wrap a piece of electrical tape around the white wire that goes to the switch to code this wire as a black hot wire. Attach the black wire from the power source to this coded white wire. Attach the white wire coming from the fan housing to the white wire from the power source. Attach the black wire coming from the fan housing to the black wire going to the switch. Connect a short piece of bare wire to the green grounding screw in the fan housing. Connect this short bare wire to the bare wires from the power source and from the switch.

2 Wiring the Switch. Install a 2½-inch switch box in the wall near the door. Use a standard single-pole switch. Wrap black tape around the white wire coming into the box. Attach this wire to one of the switch terminals and the other black wire to the other terminal. If the box is plastic, connect the bare ground wire to the green grounding screw on the switch. If the box is metal, connect a short piece of bare wire to the green grounding screw on the switch. Connect another piece of bare ground wire to the grounding screw in the box. Connect the two short bare wires to the incoming bare wire. Screw the switch into the box and install the cover plate.

3 Connecting the Fan to the Housing. Plug the fan into the motor receptacle, secure the fan into the housing and install the grille.

Wiring the Fan In-Line

If the power source is closer to the switch than the fan, run the power cable from the source into the switch. Then run another cable from the switch to the fan housing. Use cable of the same gauge as the cable that supplies power to the source.

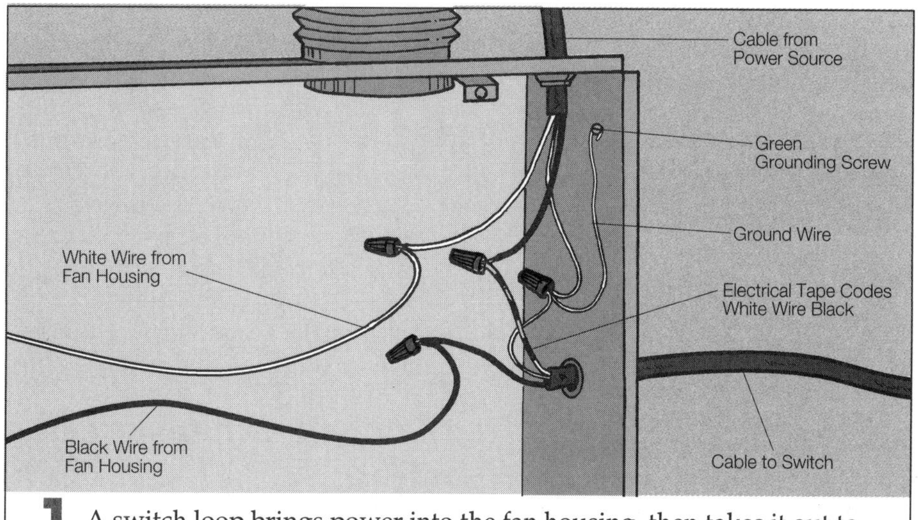

1 A switch loop brings power into the fan housing, then takes it out to a switch and then back to the housing.

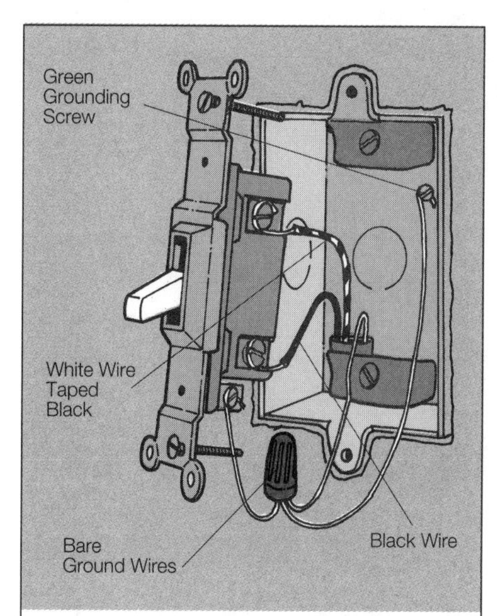

2 Connect the black wire to one switch terminal and the white wire to the other terminal. Make the grounding connections. Screw the switch into the box and install the cover plate.

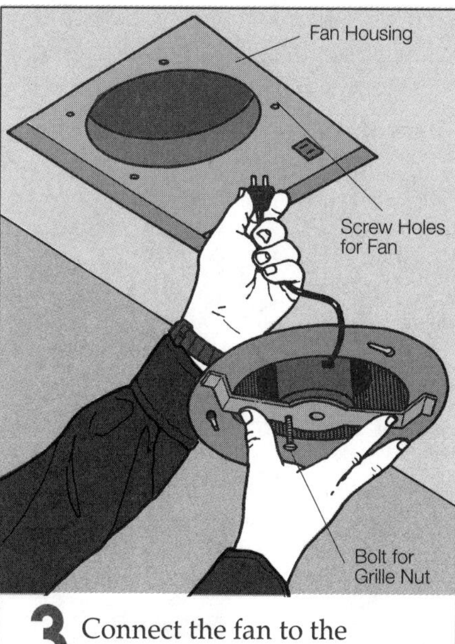

3 Connect the fan to the housing. Plug the fan motor into the receptacle and secure the fan to the housing with the screws provided. Then attach the grille.

1 Wiring the Switch. Run a length of cable from the power source into a 3½-inch switch box in the wall near the door. Run a cable out of the switch box to the fan housing. For a standard single-pole switch, connect one black wire to one switch terminal and the other black wire to the other switch terminal. Connect the white wires together. Connect a short piece of bare wire to the green grounding screw on the switch. If the box is plastic, connect this short bare wire to the two bare wires coming into the box. If the box is metal, connect another short bare wire to the grounding screw in the box. Use a wire connector to connect the two short bare wires to the two bare wires from the cable.

2 Wiring the Fan. When the fan is wired in-line, there is only one cable coming into the fan housing. Connect the black wire from the cable to the black wire from the fan housing. Connect the white wire from the cable to the white wire in the fan housing. Connect the incoming bare wire to the grounding screw in the fan housing.

Plug the fan into the motor receptacle, secure the fan into the housing and install the grille a you did for step 3 of "Wiring a Switch Loop" on page 73.

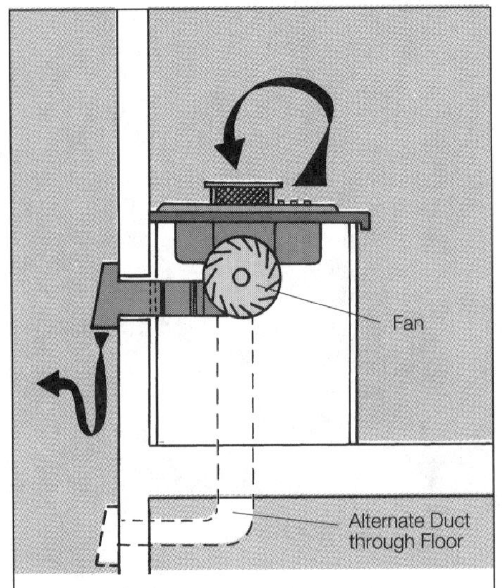

Vented Ranges. Air is pulled down through a grille in the cooking surface and discharged through a wall or floor, and ducted to the outside.

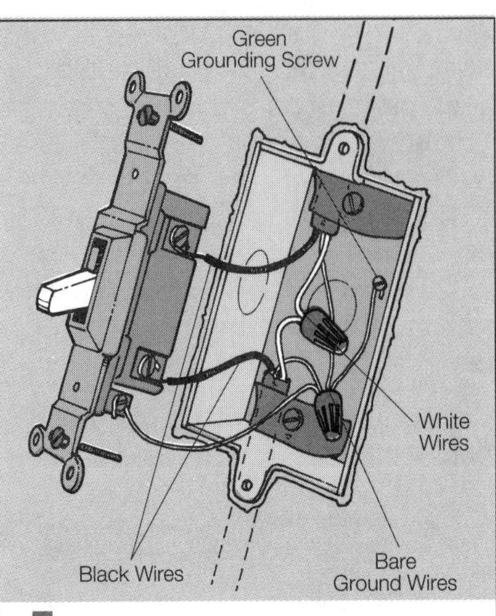

1 To wire an in-line switch, run a cable from the power source into the switch box. Run another cable from the switch box to the fan. Connect both black wires to switch terminals. Connect white wires to each other. Connect bare wires to the grounding screw and to each other.

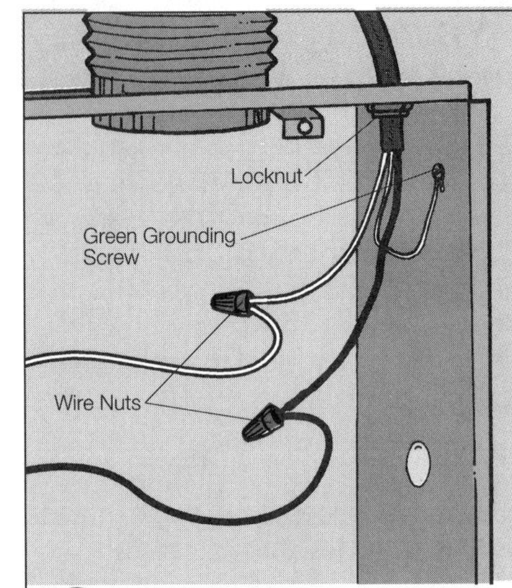

2 Connect the white cable wire to the white fan wire. Connect the black cable wire to the black fan wire. Connect the bare cable wire to the green grounding screw in the fan housing.

Kitchen Exhaust Options

Contaminated air that results from cooking can be exhausted by a fan mounted in the range itself or by a separate range hood mounted above the appliance.

Vented Ranges. Some ranges contain their own exhaust systems. Fans under the cooking surface draw in foul air and expel it to the outside through a nearby wall or a duct in the floor.

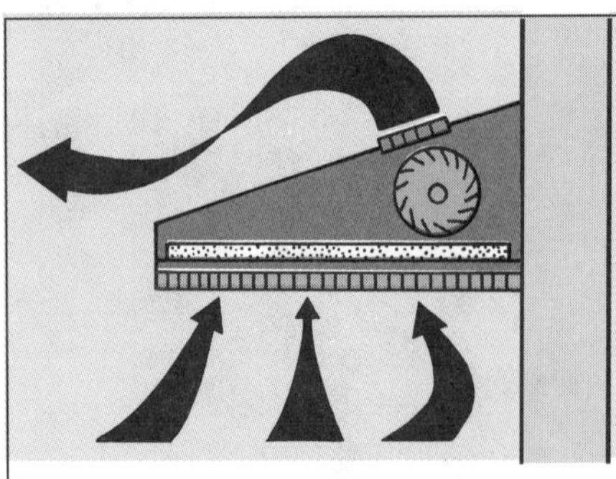

Nonvented Range Hoods. Room air is drawn through a grille and filter, then recirculated to the room.

Nonvented Range Hoods. Nonvented (recirculating) systems pull room air through a filter and return it to the room through a grille. Because they remove only some of the contaminants and none of the humidity, they are used only in cases in which it is impossible to vent to the outside.

Vented Range Hoods. Vented range hoods draw air through a filter located in the bottom of the range hood, and exhaust it outside the house via a duct in the wall or roof.

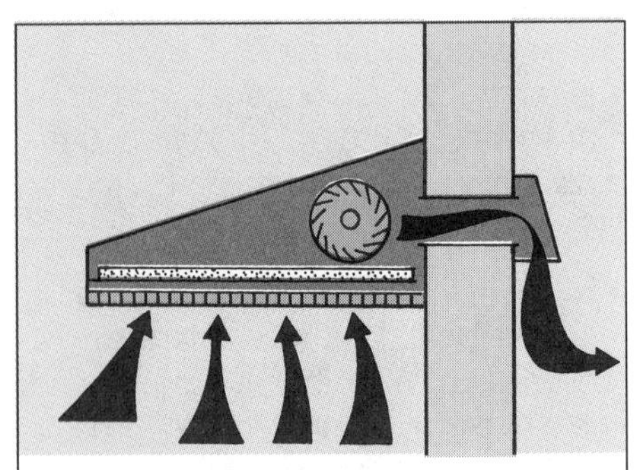

Vented Range Hoods. These hoods exhaust stale air to the outside.

Selecting a Range Hood

The size of the range hood will be determined by the place in which it will be installed. If it is going to be installed above a freestanding island, for example, the range hood can be any size you like as long as it is larger than the cooking surface. The broader and deeper the hood, the better. Wall mounted hoods work best when they overlap the cooking surface by at least 3 inches on each side. Standard range hoods are available in widths of 30, 36, 42 and 48 inches. Hoods can accommodate rectangular or round duct-work. With the proper adapter, a range hood can be joined to a duct of almost any size and shape, but the simplest way is to match the duct to the exit opening on the hood.

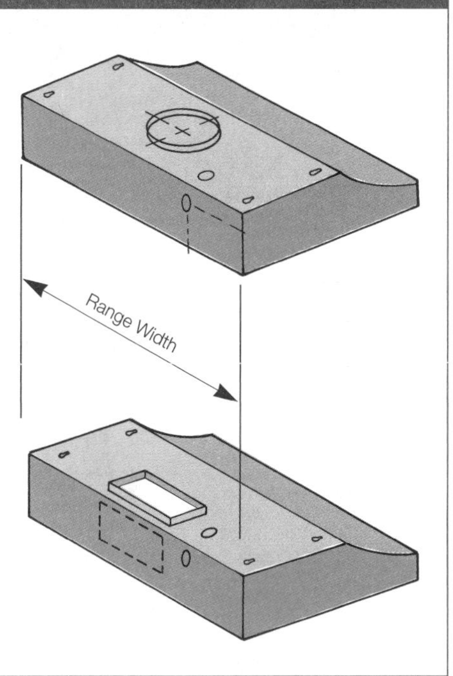

Planning the Layout

The hood will be mounted 24 to 36 inches above the cooking surface. Plan the duct layout to obtain the shortest path to the outside. If the range sits against an outside wall, vent the hood directly through the wall. A range on an interior wall may be vented via a duct through the nearest outside wall or through the roof or soffit. Select the appropriate termination cap for the wall or roof, then determine the type and length of ducting required. Avoid having to install elbows by using 4- or 6-inch-diameter flexible duct.

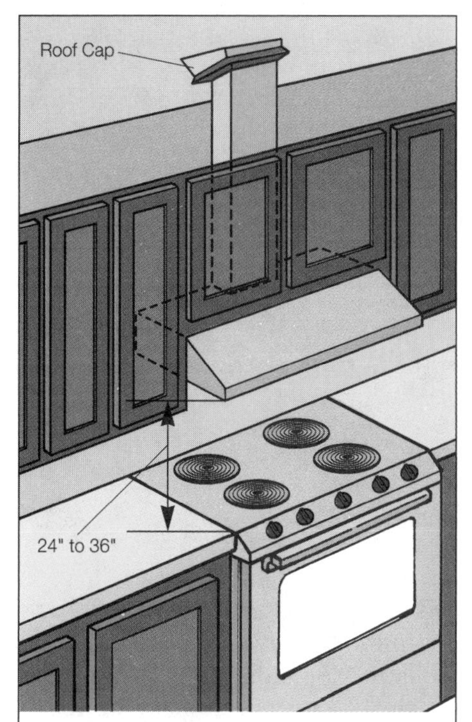

Planning the Layout. Range hoods can exhaust through a wall or roof. Plan to mount the hood between 24 and 36 in. above the cooking surface.

Installing a Range Hood

The steps below describe how to install a vented range hood that is connected by duct-work to the roof or an exterior wall. The installation of a hood on an exterior wall uses the same general method, minus the duct. A nonvented range hood is simply mounted above the range.

1 Cutting the Hole. The center of the hole will be located between wall studs or ceiling framing. Make sure it is not in the line of wiring or piping (an electronic stud finder can be helpful). Use a compass to mark the cutout which will be an inch larger than the sleeve of the termination cap. From the inside, drill a small hole completely through the center using a long bit (8 inches or more). Drill a larger starter hole along the cut line and use a keyhole saw or saber saw to cut out the inside wall finish. Remove the insulation in the area of the cutout. Then, working from the outside, repeat the process for the outer wall.

2 Installing the Duct-Work. Assemble the duct parts and seal the joints with duct tape. If the duct-work runs to a remote wall it can be hidden behind an enclosure (such as a dropped ceiling or a soffit over kitchen cabinets).

Caution: *Wear heavy work gloves when working with sheet metal.*

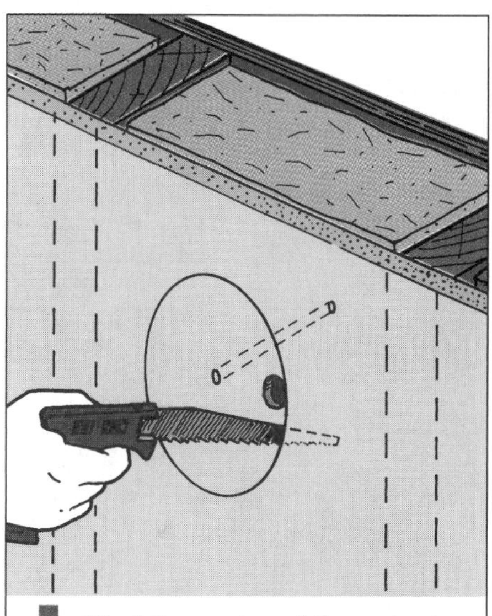

1 Find the center of the cutout and drill through the wall to locate the center on the outer surface. Cut out the inside piece and remove insulation (if any). Then repeat the process for the outer wall, working from the outside.

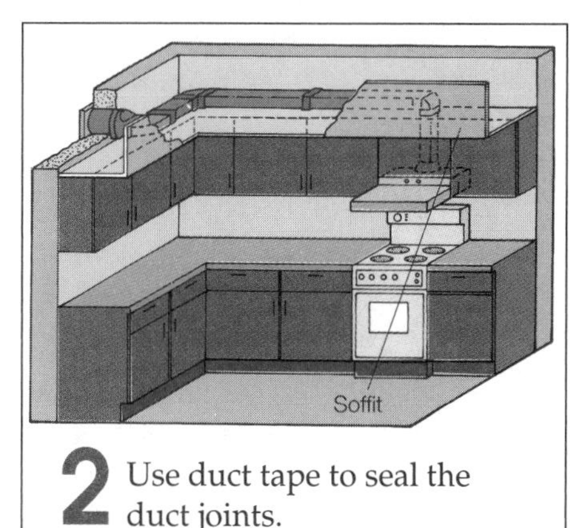

2 Use duct tape to seal the duct joints.

3 Install the termination cap and accessories. Use a flexible exterior caulk to seal around exterior openings.

4 Use the dimensions that come with the hood to lay out the openings and attachment points on the underside of the cabinetry.

3 Capping off the Outside. Insert the termination cap from the outside. Caulk the flange that contacts the wall or roof (the soffit vent does not have to be caulked). Use butyl rubber or acrylic latex caulk for walls; use asphalt caulk for roofing.

4 Installing the Hood. Installation methods vary from model to model, so use the manufacturer's instructions that come with the hood as a guide for cutting the duct opening, drilling holes, securing the hood and completing the wiring.

Ventilation for Cooling

When air moves over a person's skin, the person feels cooler. Moving air can make a room feel as comfortable at 85 degrees Fahrenheit as it would in still air at 78 degrees Fahrenheit. To be effective, the air must move at a speed of between 100 and 300 feet per minute (1 and 3 miles per hour). The amount of cooling provided by natural ventilation depends not only on how fast the air moves, but also on its temperature, humidity and the clothes a person is wearing.

The simplest way to provide ventilation is with windows that are sized and located to capture outside breezes. Room fans provide portable ventilation for part of a room, while whole-house fans ventilate several rooms.

Installing a Whole-House Fan

A whole-house fan cools the house by moving large quantities of air through the living space and exhausting it through the attic. It can make a house more comfortable during some parts of the year and reduce the dependence on air-conditioning. Though it will not cool as effectively as an air conditioner, a whole-house fan is far cheaper to install and costs one-fifth to one-tenth as much to operate.

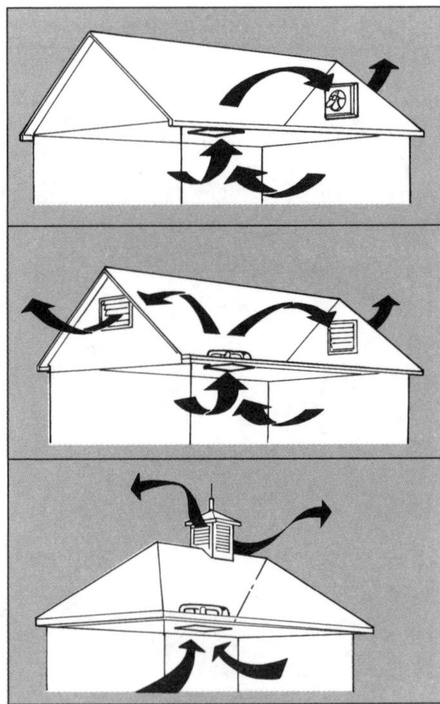

Installing a Whole-House Fan. Mount the fan in the gable end of a pitched roof and draw air through a grille in the attic floor, or vice versa. For houses that have hip roofs, fans mount in the attic floor and exhaust air through a cupola at the roof peak.

A fan mounted in a central location such as a hall ceiling, pulls air into the house through partially-opened windows and exhausts it through grilles in the gable ends of the attic or vents on the roof. In an alternate arrangement a fan can be mounted in a gable-end window or in the gable itself. With this arrangement, the fan pulls air through a grille in the ceiling, or the attic hatch simply is left open.

Whole-house fans are available in diameters of 24, 30, 36 and 42 inches. A 24-inch fan suits a total floor area of around 1,900 square feet; a 30-inch fan handles an area of 2,500 square feet; a 36-inch fan works for an area of 3,300 square feet. When selecting a fan, check the free area required for exhaust and compare this figure to the total area of gable or roof vents in the attic. If the vent area is not stated, figure 1 square foot of vent area for each 750 cfm of fan capacity. Fans can be switched manually, hooked to a timer switch, or controlled automatically by a thermostat. The following shows how to install a whole-house fan in the attic floor.

1 **Cutting the Ceiling.** Drill a pilot hole in the ceiling at the center of the location for the fan. Then poke a wire through to mark the position. From the attic, remove insulation from the general area. Cut an opening in the ceiling the same size as the fan shutter. Remove the ceiling drywall (or lath and plaster) and cut the ceiling joists 1½ inches back from the edges.

2 **Framing the Opening.** Nail header joists to the cut ends of the ceiling joists, then nail in trimmer (side) joists. Double the headers if they span more than two ceiling joists. Make a fan support platform of 1x6s to fit around the top of the opening. Nail the platform to the framing.

3 **Mounting the Fan.** Set the fan in place and screw it to the platform. Run a power cable to the fan from a nearby junction box and wire the fan and switch according to the instructions that come with the fan.

1 Cut out an opening in the ceiling to the size required by the fan. Cut joists back 1½ in. to allow for framing the opening.

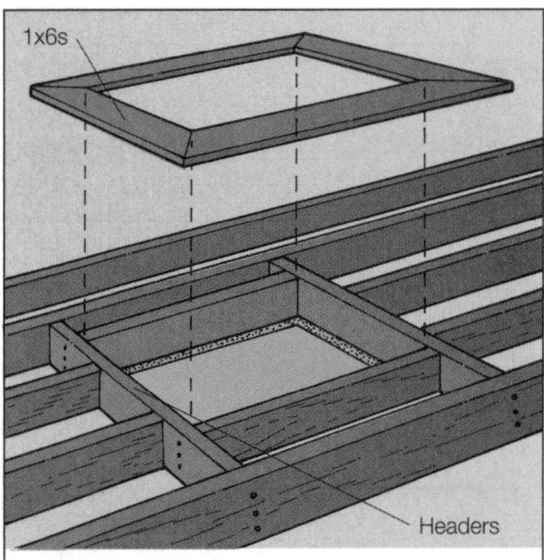

2 Nail header joists to the joist ends. Add trimmer joists. Then make a platform out of 1x6s and nail over the top of the opening.

Install the shutter from below by driving wood screws into the framing.

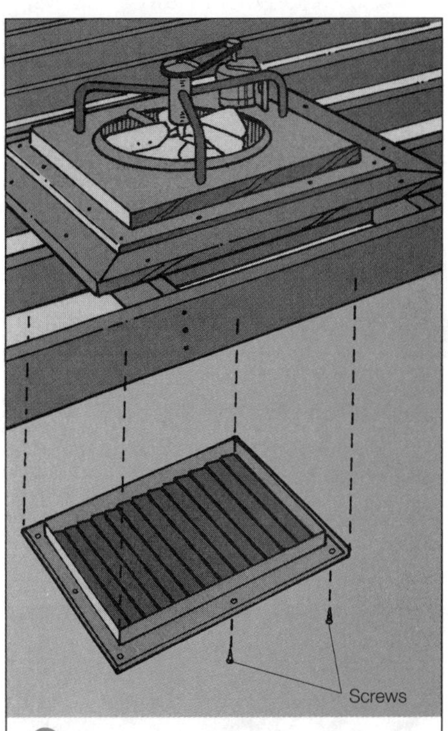

3 Secure the fan over the support platform, install the wiring and attach the grille on the ceiling below.

Replacing a Light Fixture With a Ceiling Fan

Like a whole-house fan, a ceiling fan is a lower-cost choice for cooling than is air-conditioning. A ceiling fan can cool a single room simply by keeping the air moving. In winter, a fan mounted in a high ceiling enhances comfort by pulling warm air to the floor. Ceiling fans are available in a wide assortment of fixtures from basic fan units to fans that contain built-in light fixtures. Before purchasing a new fan, make sure the proposed location offers ample space. There must be at least 24 inches between the tip of the fan blade and a nearby wall, and the bottom of the fixture must clear the floor by at least 7 feet.

Usually a light fixture can be replaced by a ceiling fan entirely from inside the room. Codes prohibit hanging the fan from the electrical box however, so a separate support cable that is capable of holding up the fan will have to be used (some fans weigh as much as 30 pounds). Support brackets are available in kits sold separately from the fan fixture.

1 **Removing the Old Fixture and Box.** First turn off the power to the light fixture at the electrical panel. Remove the fixture and disconnect the wires. To identify the two wires that were connected to the light fixture, put a piece of tape over them. Remove the box by unscrewing the center fastener and cable clamps. Temporarily displace insulation found near the box and bracket. (For

blanket insulation, use a stick or wire to prop up a tent over the area. The tent can be pulled down after the new bracket is in place.) Use a mini hack-saw to sever the old support bar and then bend the bar out of the way.

2 Installing the New Hanger Brace Bar. Feed the new brace bar into the cavity along the direction of the joists. Then rotate it 90 degrees so that it spans crosswise in the cavity and the feet rest on the ceiling drywall. Turn the bar on its axis, allowing it to unscrew and extend until the legs are snug to the joists. Before tightening it make sure the bracket is centered over the hole. (Since you will not be able to see into the hole, these last two actions have to be done by feel only. Be patient and take your time).

3 Attaching the New Box. Remove knock-out tabs from two holes in the new box. Attach wire clamps to the holes. Feed one cable through each clamp and tighten the clamps inside the box to secure them. Next, refer to manufacturer's instructions to attach the new hanger assembly to the bracket and box.

4 Attaching the Fan Hanger Bracket. Mount the fan hanger bracket on the junction box according to the manufacturer's instructions. Pull down the wires through the bracket.

5 Connecting the Fan. Follow manufacturer's instructions for mounting the fan onto the bracket and connecting the wiring to the appropriate fan and light wire leads. Finally, attach the light or cover plate.

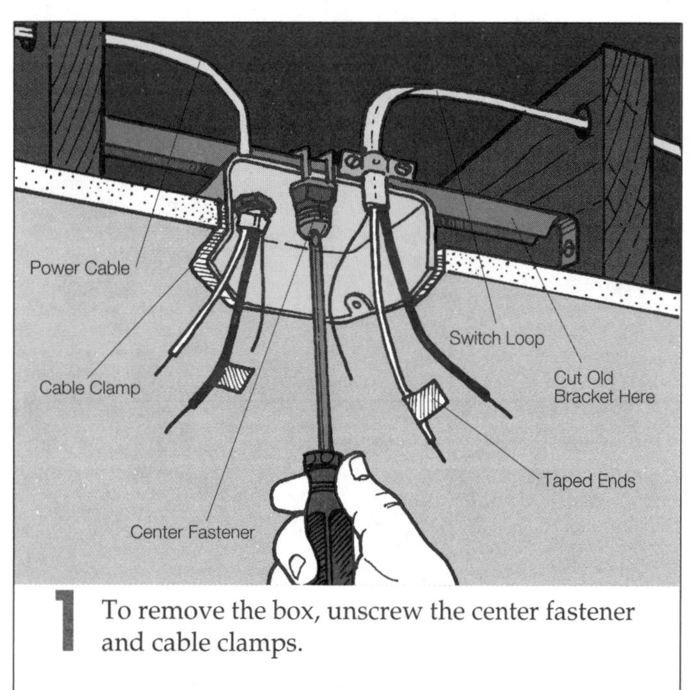

1 To remove the box, unscrew the center fastener and cable clamps.

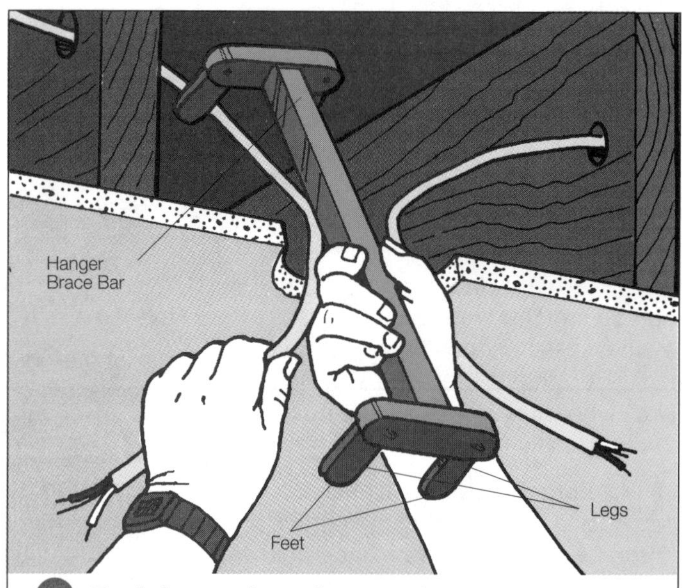

2 Feed the new brace bar into the cavity along the direction of the joists, then rotate it 90 degrees so that it spans crosswise in the cavity and the feet rest on the ceiling drywall.

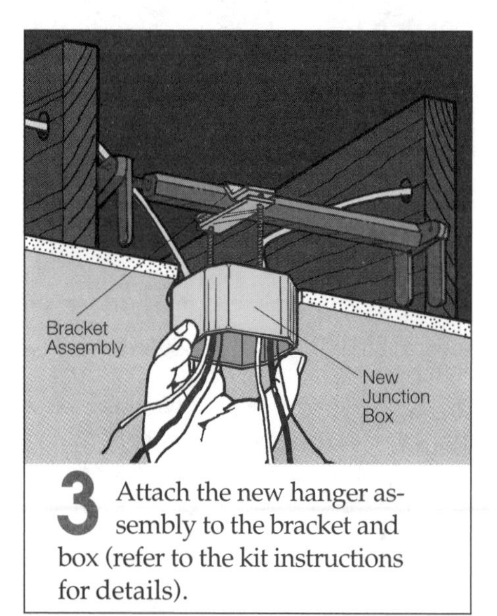

3 Attach the new hanger assembly to the bracket and box (refer to the kit instructions for details).

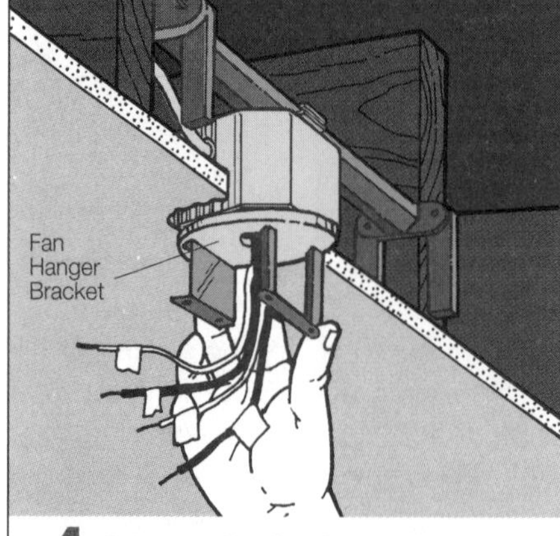

4 Mount the fan hanger bracket on the junction box according to manufacturer's instructions. Pull the wires through the bracket.

5 After the fan is mounted, attach the light or cover plate.

G L O S S A R Y

Air Barrier Tape A moisture-proof tape that covers the joint and sticks to the substrate. Duct tape or contractor's tape work well.

Batt Insulation Pre-cut lengths of flexible insulation, such as fiberglass and mineral wool, that come packaged in several pieces.

Blanket Insulation Flexible insulation, such as fiberglass or mineral wool, that comes packaged in long single rolls.

Blown Insulation Cellulose and fiberglass loose-fill insulation blown into an attic or existing wall using pneumatic equipment.

Cellulose (loose-fill) Recycled newspaper that has been treated with a fire-retardant chemical and is used as insulation.

Conduction The flow of energy through a solid. Conductive heat is transferred from a warmer surface, such as a radiator, to a colder one, such as your hand.

Convection The transmission of a heat by liquid or gas. Convective heat is felt as warm air wafting above a hot source, such as a candle flame.

Door Sweep Weatherstripping that mounts to the bottom of the door. It consists of an extruded aluminum strip that holds a flexible vinyl strip.

Double-Glazed Window A window consisting of two panes of glass separated by a space that contains air or argon gas. The space provides most of the insulation.

Drywall Also known as wallboard, gypsum board, plasterboard, and by the trade name Sheetrock; a paper-covered sandwich of gypsum plaster used for wall and ceiling surfacing.

Eaves Lower edge of a sloped roof that projects over the outside wall.

Foam in a Can Polyurethane foam sealant packaged in a pressure can. Used to fill irregular, hard-to-reach cracks and gaps that cannot be plugged effectively with caulks and gaskets.

Green Wire In a cable, the wire that functions as a ground wire.

Insulation Baffles 1/2-inch-thick polystryene beadboard. These devices are formed into a corrugated shape that guarantees air channels next to the roof.

Insulation Facings A foil or kraft paper facing that covers blanket, batt, or rigid insulation. Foil-faced can be used as a vapor barrier; kraft paper is only a marginal vapor barrier.

Jamb The inside face of a window or door.

Joist One in a series of parallel framing members that supports a floor or ceiling load. Joists are supported by beams or bearing walls.

Loose-fill Insulation Bags or pellets of insulation that can be poured where needed or blown-in by pneumatic equipment.

On Center A point of reference for measuring. For example, "16 inches on center" means 16 inches from the center of one framing member to the center of the next.

Polyethylene Sheet A material well suited to retard vapor passage in a floor, wall or ceiling. Comes in thicknesses of 4, 6 and 8 mils.

R-value A number assigned to thermal insulation to measure the insulation's resistance to heat flow. The higher the number, the better the insulation.

Radiant Barrier Materials Highly reflective surfaces that block radiant heat energy. Available in the form of rigid sheets (foil-faced rigid foam or plywood sheathing) and rolls of flexible sheeting.

Radiation The emission of energy from an object. Heat waves from the object radiate to cooler objects. Like radio waves, this form of energy passes through the air without heating it. It becomes heat only after it strikes and is absorbed by a dense material.

Rafter A structural member that supports the sloping roof of a structure.

Rake The part of the roof that projects over the gable ends.

Ridge The horizontal line at which two roof planes meet when both roof planes slope down from that line.

Ridge Vent A vent that is made of various materials and covers the ridge of the roof. Ridge vents may be packaged in rigid sections or rolls.

Rigid Insulation Boards of insulation that are composed of various types of plastics. Rigid insulation offers the highest R-value per inch of thickness.

Sash The framework into which window glass is set. Double-hung windows have an upper and lower sash.

Sill Plate Also known as a bottom plate; a horizontal framing member attached to the subflooring that supports the wall studs.

Single-Glazed Window A window with a single pane of glass held in place with glazing putty; single glazing is not as energy-efficient as double glazing.

Sleepers Boards laid over a concrete floor as a foundation for the subflooring of a new floor.

Soffit The finished underside area of an eaves.

Soffit Vent Vents placed in the soffit (the most common way to let air into the roof). Example: ventilator plugs, continuous soffit strips vents, and rectangular vents.

Spring-Metal Weatherstripping that is made of copper strips formed into wide V-shaped mounts. Spring-metal is installed inside the window track or door jamb.

Stops Strips of wood nailed to the head and side jambs to prevent a door from swinging too far when it closes. Also to keep window sash in line.

Stud Vertical member of a frame wall, placed at both ends and usually every 16 inches on center. Provides structural framing and facilitates covering with drywall or plywood.

Subfloor The floor surface below a finished floor. Usually made of sheet material such as plywood; in older houses it is likely to consist of diagonally attached boards.

Switch Loop Installation in which a ceiling fixture is installed between a power source and a switch. The power passes through the fixture box to the switch. The switch then sends power to the fixture itself.

Tiger's Teeth Wire supports used to secure unfaced blankets or batts between rafters.

Toenail Joining two boards together by nailing at an angle through the end, or toe, on one board and into the face of another.

Top Plate Horizontal framing member, usually a 2x4, that forms the top of a wall. Attaches to the tops of wall studs and supports floor joists and rafters.

Vapor Barrier Material used to block out the flow of moisture.

INDEX